Overton Park

Neither nature nor people alone can produce human sustenance, but only the two together, culturally wedded.

—WENDELL BERRY—

The Unsettling of America

BROOKS LAMB

OVERTON PARK A People's History

THE UNIVERSITY OF TENNESSEE PRESS
Knoxville

COPYRIGHT © 2019 BY THE UNIVERSITY
OF TENNESSEE PRESS / KNOXVILLE.

ALL RIGHTS RESERVED. MANUFACTURED
IN THE UNITED STATES OF AMERICA.
FIRST EDITION.

*LIBRARY OF CONGRESS CATALOGING-
IN-PUBLICATION DATA*

NAMES: LAMB, BROOKS, AUTHOR.
TITLE: OVERTON PARK: A PEOPLE'S HISTORY /
BROOKS LAMB.
DESCRIPTION: KNOXVILLE: THE UNIVERSITY
OF TENNESSEE PRESS, [2019] | INCLUDES
BIBLIOGRAPHICAL REFERENCES AND INDEX. |
IDENTIFIERS: LCCN 2018013201 (PRINT) |
LCCN 2018014560 (EBOOK) | ISBN
9781621904618 (PDF) | ISBN 9781621904625
(KINDLE) | ISBN 9781621904601 (PBK.)
SUBJECTS: LCSH: OVERTON PARK
(MEMPHIS, TENN.)-HISTORY. |
PARKS-TENNESSEE-MEMPHIS-HISTORY.
CLASSIFICATION: LCC SB482.A52 (EBOOK) | LCC
SB482.A52 L36 2019 (PRINT) |
DDC 363.6/80976819-DC23

LC RECORD AVAILABLE AT
HTTPS://LCCN.LOC.GOV/2018013201

CONTENTS

ILLUSTRATIONS

ACKNOWLEDGMENTS

Overton Park is special because of the many people who have lent it their time, effort, and love. The same can be said of this book. Over several years, many mentors, professors, friends, and family members offered advice, edits, and support. This book would not exist without their help. I will inevitably leave someone out, and for that, I apologize. That being said, I extend my deepest gratitude to the following people.

First, I would like to thank the University of Tennessee Press for believing in this book. I especially want to thank Thomas Wells, who showed grace and kindness to a young, somewhat clueless author. I would also like to thank Gene Adair, Scot Danforth, Linsey Perry, Tom Post, and Jon Boggs for their invaluable assistance.

The heart and soul of this book lie in the many people who took the time to share their memories of Overton Park. A very special thanks goes to Anne Pitts, Blanchard Tual, Charlie Newman, Donnie Bailey, Fred Davis, Gigi Wischmeyer, James Jalenak, Janet Hooks, Jimmy Ogle, Johnnie Turner, Kathy Fay, Martha Kelly, Melanie White, Mike Cody, Richard Meek, Sally Jones Heinz, Steve Cohen, and Willy Bearden. Some of your stories made me laugh. Others made me cry. All deepened my understanding of Overton Park and its significance. Additional thanks must go to Willy Bearden, who shared hundreds of historic photos of Overton Park.

The staff at Overton Park Conservancy offered endless support, encouragement, and assistance. I'm grateful to Andrew Weda, Gina Christian, Rose Baker, and Sheila Holbrook-White. Special thanks are due as well to the following current and former staff members. To Susannah Barton, I owe thanks for sharing an office with an intern (and never once complaining aloud about it). Eric Bridges taught me about the Old Forest; his knowledge of and passion for trees and ecology are inspiring. Tina Sullivan's leadership exemplifies persistence and dedication. And Melissa McMasters gets a special thanks for being my internship supervisor of three years. She encouraged me to do this project. She read the manuscript and offered thoughtful feedback, showing constant support and kindness, even in the most difficult of times. For all she has done, and, most important, for her friendship, I thank her.

To the faculty and staff at Rhodes College, I offer appreciation. Through technical support, writing advice, and countless conversations both in and outside the classroom, the following people directly and indirectly left their imprint on this book: Bill Short, Bill Troutt, Charlie Kenny, Courtney Collins, Erin Dolgoy, Judith Haas, Michael Drompp, Michael Nelson, Rebecca Tuvel, Sarah Boyle, Tait Keller, and Victor Coonin. I'm also grateful to Shannon Hoffman and the Rhodes Col-

lege Bonner Program, as well as Charles McKinney and the
Rhodes Institute for Regional Studies, for financial, emotional,
and intellectual support.

A few professors-turned-mentors played significant roles
in this book, investing their time and effort. Their names are
only mentioned here, but their influence is felt on each and
every page. Thanks to Charles Hughes for his thoughtful advice and editing. Sipping coffee, we pored over his handwritten
edits, searching for ways to improve the book. The revisions
were well worth the time, especially after seeing his reaction
to news of the book's publication. Thanks to Tim Huebner for
his thorough reading of the manuscript and for his advice on
publication. Most important, I thank him for his constant support—in person, on the phone, and via email. Knowing that
he believed in me made me believe in myself. And thanks to
Robert Saxe for all that he did to make this book possible. He
read multiple drafts of the manuscript, offering both specific
editing and stylistic suggestions. He leapt outside his comfort
zone to discuss environmental history and land ethics in a
one-person class we designed together. He talked me through
struggles, invited me to dinner at his home, and even let me
lie down on his office floor when I wasn't feeling well. With
selflessness and wit, he provided indispensable guidance.

Educators outside of Rhodes were helpful, too. David Stirk
of Princeton University kindly sent park-centered reading
suggestions, and Dr. Charles Crawford of the University of

Memphis offered valuable advice about oral history. Other people in the Memphis community, including Baird Callicott, Carl Awsumb, George Cates, Jeanne Jemison, and Jim Lanier, also gave valuable insights. Two members of the Memphis community proved instrumental. Through his work in conservation, Cary Fowler offered inspiration. He is a kind, accessible, and helpful mentor. Bill Haltom has also become an important mentor, offering feedback on the manuscript, helping me think through publication and, in the process, becoming a close friend. He also bought me several lunches at the Little Tea Shop in Memphis, for which I still owe him. Knowing that he is only a phone call away gives me great comfort.

Several other authors were key to this book's publication. Wayne Dowdy and Otis Sanford, both of whom reviewed the manuscript at UT Press's request, offered thorough, detailed suggestions for improving the book. Their edits are deeply appreciated. Other authors helped in the planning process. Varina Willse generously spent time with me on the phone, discussing her approach to writing oral history. And although we've never met, I'm grateful to Wendell Berry for helping me think about the importance of place. His concepts of imagination and affection, in particular, had a tremendous impact on this book.

Friends, at Rhodes College and beyond, were also instrumental to this book's success. They read drafts, gave support, and—most important—willingly endured and perhaps enjoyed dozens of conversations about Overton Park. To the following people, I offer thanks: Alexis Moore, Bonnie Whitehouse, Emily Watkins, Henry Smith, Kirkwood Vangeli, Lee Sands, Natasha Main, Omair Arain, and Zain Virk. I also appreciate fellow Rhodes Bonner Scholars, as well as fellow Truman Scholars, for both challenging and encouraging me. A few friends deserve special recognition. For being the first to read the manuscript and offer feedback, I owe Maddie McGrady a huge debt. Without her, I might have given up on this project. For her constant support and for helping me with interview transcriptions, I'm grateful to Jolie Grace Wareham. At times, she quite literally dropped what she was doing to help with this project. For proofreading the manuscript, engaging in countless Overton Park–focused discussions and service activities, and offering endless encouragement, I thank Regan Adolph. Her selfless love and commitment made the completion of this book possible.

Finally, I thank my parents, Ken and Angela, and my brothers, Patrick and Michael. My parents taught me to work hard and to care for others. Their example is one that I can only hope to emulate. Patrick has shown what dedication and determination look like in action. He is a loving father and a loyal brother. And Michael, in addition to editing the manuscript, plays the roles of both my most-trusted mentor and friend. This book is better because of him. For my family's love, support, and effort, I'm grateful.

INTRODUCTION

Overton Park is the heart of Memphis. It's where nature enthusiasts escape into the Old Forest, where families visit the Memphis Zoo, and where art lovers flock to view famous paintings in the Brooks Museum. It's where musicians awe crowds at the Levitt Shell, where students hone their skills at the Memphis College of Art, and where golfers play the game they love on the Links. Runners and walkers visit daily. Parents and their children do, too. The park has something for everyone.

These present-day aspects of the park make it a favorite among locals and visitors alike. But what can their stories, the memories they have collected over the years, tell us about

Overton Park's history? As it turns out, the park's past is even more dynamic than its present. Johnnie Turner remembers being arrested and jailed when working to desegregate the Overton Park Shell. Attending a church service there, she had just placed a dollar in the offering plate when she was escorted outside by the police. Michael Cody reflects on a controversial antiwar speech he gave in the park while the United States was fighting in Vietnam. As he gave his speech, support for and criticism of his position simultaneously filled the air. Charlie Newman recalls arguing in the U.S. Supreme Court to save Overton Park from an expressway. Had he and other determined citizen-stewards not waged this decades-long battle, the tulip poplars and towering oaks in the Old Forest would have been replaced with speeding automobiles on asphalt. These memories show that Overton Park is a refuge whose history speaks to the changing social and physical landscape of American life, a microcosm of the movements and moments that have shaped our nation.

For over a century, the park has, in one way or another, played host to important historical events. The Great Depression touched the park through the Overton Park Shell, now renamed the Levitt Shell, a Works Progress Administration project to employ struggling Memphians. The conflicts of World War II invaded this space when the Japanese Garden—once a hallmark of the park—was vandalized and then officially removed following the attack on Pearl Harbor.

The civil rights movement transformed the park forever when courageous college students dared to fight racial oppression by conducting sit-ins. Their efforts helped make the park a place for all Memphians regardless of race. Following the civil rights movement, another source of grassroots activism entered the park: the environmental movement. Less than a year after the first Earth Day was observed on April 22, 1970, Citizens to Preserve Overton Park secured a landmark victory in the Supreme Court to halt an interstate from destroying the Old Forest, establishing a precedent that would be utilized by environmentalists for generations.

Other moments in the park also reflect larger trends in American history. Elvis Presley, who defined a new popular culture and heralded the arrival of rock and roll, performed his first paid concert in Overton Park. And from commemorating America's sacrifices in battle at Veterans Plaza to hosting anti–Vietnam War protests near Rainbow Lake, the park has witnessed patriotism and pacifism. In so many different ways, Overton Park gives us a lens to examine our past, understand our present, and determine our future.

These stories and moments—from dramatic events like sit-ins and antiwar speeches to everyday occurrences like walks through the Old Forest and visits to the zoo—prove that Overton Park belongs to the people. Without patrons, Overton Park's history would be empty. And because Memphians make the park's history, they are best positioned to tell its story. Oral history allows this to happen, enabling us to hear the story of the park in the most personal way possible. This approach doesn't rely on dry notes from the minutes of a meeting, and it doesn't limit itself to what can fit in one section of a newspaper. Instead, it offers us the lived experiences of people who populated and protected the park. If we are to truly understand the importance of this place, then we must hear from the people who have witnessed the park's impact on both individuals and the community.

History is often devoid of emotion, and rightfully so. We historians strive to be objective, even if it is an impossible task. But oral history affords an insight into what Memphians have experienced throughout the park's history—what they thought, felt, and saw. Studs Terkel, one of America's most famous oral historians, once said, "People are hungry for stories. It's part of our very being. Storytelling is a form of history, of immortality, too. It goes from one generation to another." By telling the story of Overton Park through oral history, we ensure that it won't be forgotten.

Through conversations with an eclectic and diverse group of Memphians, this book offers a personal history of the park itself, highlighting times of peace and conflict and the ways in which citizens responded to adversity to defend the park they loved. As important as the buildings and spaces within the park have been, history never would have been made without the Memphians who made it. A concert stage is only a slab of concrete without performers or fans. An old-growth forest sits unexplored until a curious mind wanders through it. A playground lingers as a pile of metal and wood until a child brings forward her imagination. It takes a people to make a park. This is their story.

It is also mine. I came across the park only after an intense bout with homesickness during my first year at Rhodes College. Having grown up on a small farm in rural Tennessee, I longed to hear cows and crickets instead of automobiles and airplanes. Thankfully, service provided a system of support. After transformative experiences in community gardens and with The Land Trust for Tennessee, I began an internship with Overton Park Conservancy that enabled me to strengthen my passion for environmental issues. Though I was unsure of what to expect from working in a city park, I soon realized that the position was a perfect fit. I shared the conservancy's mission, developed friendships with the staff members, and uncovered a love for the 342-acre park.

For three years, I volunteered with the conservancy for ten hours each week. During this time, we created a new membership program, dug a new drainage system in the dog park, and picked up trash in the Old Forest. We planted trees, spread mulch on playgrounds, and removed invasive weeds. We even raised $1 million in a two-month period to protect Overton Park's Greensward, sometimes standing out in the rain with buckets and hand-drawn posters, hoping for—and often receiving—support from passersby. In countless ways, we worked with others in the community to improve the park, each of us sharing a vision to create an even more welcoming, entertaining, and beautiful place.

The idea for this book arose from interactions with others. While planting trees or removing litter from the trails, other Memphians casually shared their own personal stories about the park. Collecting these memories and weaving them together to tell Overton Park's history, I realized, could lift up their stories and show the importance of this place to a wider audience. Armed with support, advice, and a voice recorder from my friends at the conservancy, I spent months interviewing Memphians with unique experiences in Overton Park.

Honestly, I didn't expect to uncover so many moving memories. Congressman Steve Cohen remembers playing baseball in the park as a child. Kathy Fay and Richard Meek recall falling in love—both with each other and the animals—while working as zookeepers in the Memphis Zoo. Willy Bearden reflects on two decades of Thanksgiving morning walks through the park with his family and closest friends. These intimate and moving stories, among dozens of others, reinforce the idea that it takes a people to make a park.

Despite the merits of using personal memories as the main sources for this book, shortcomings loom. Many of the stories from the park's early years are lost to time. Moreover, despite deliberate attempts to interview people from differing backgrounds, the stories here do not reflect the exhaustive diversity of the people who have made the park's history. For this, I apologize. Nonetheless, the memories cited in this book offer powerful accounts of Overton Park's story, accounts that can help us understand this space's significance. Through the eyes of others, readers can live some of the park's signature moments. The memories cited in this book provide a unique and personal glimpse into the park's past, a glimpse that brings Overton Park's history to life.

Throughout this project, different people from different eras have shared different stories about the park, but one theme has been persistent: Overton Park is *our* park. It belongs to the people. And that's why Memphians should tell its story.

CHAPTER 1

EARLY MEMORIES

The 1940s and 1950s

Founded in 1901, Overton Park began as an attempt to connect Memphians with the land. This space, said park designer George Kessler, is "a heritage to the public for the enjoyment of nature."[1] Creating what is termed a "pleasure ground" by park historians, Kessler designed walkways that meandered throughout the park, carriage paths that wound along the perimeter and through the forest, and green spaces that highlighted the natural environment in an attempt to bring the countryside to the city. White Memphians—until 1963, the park and all its facilities were strictly segregated—embraced Kessler's design and often visited Overton Park to escape the chaos of the city. As Kessler had planned, this

George Edward
Kessler (July 16,
1862–March 20,
1923). Photo cour-
tesy of William
Bearden, author of
*Images of America:
Overton Park.*

space was initially valued for its simplicity, its commitment
to remaining a nature haven.

Much to the chagrin of Kessler, however, Overton Park
soon evolved into a vibrant social and cultural destination.
Beginning with the implementation of the Overton Park golf
course and the Memphis Zoo in 1906, the park's focus shifted
away from solely providing an environmental oasis to creating
a more entertaining experience. Park institutions like the zoo
and the golf course provided new opportunities for recreation,
education, and leisure, as did the playground that was built in

1911 and the Japanese Garden that was constructed in 1914. The transformation of the park's character continued with the opening of the Brooks Memorial Art Gallery in 1916, the Overton Park Shell in 1936, and the Memphis Academy of Art in 1959.[2] These three institutions—all focused on developing the creative capacities of Memphians—highlighted the importance of music and art in the city, granting the park a cultural significance. With these new amenities, Overton Park was gradually becoming a destination for Memphians of all ages and interests.

BEFORE THE MEMORIES— AND BEFORE OVERTON PARK

Unfortunately, the personal memories of Overton Park's first few decades are lost to time. Firsthand accounts of carriage rides through the forest, the Brooks's opening, or the Shell's first performance would offer an intimate glimpse into this era of Overton Park's history. But by briefly studying the park's early years with a more conventional historical approach, we can build a foundation that enables us to progress into the eras that are still remembered. Before we can study Overton Park itself, however, we must first understand the larger American parks movement.

The urban parks effort in the United States began in New York City with Frederick Law Olmsted, America's most famous landscape architect. Olmsted—who studied agricultural science at Yale and farmed for several years, thus establishing an agrarian disposition—designed the nation's first urban park during the mid-nineteenth century, providing a natural haven within the bustling city.[3] New York's Central Park gave life to the idea of city dwellers escaping mayhem and entering tranquility, finding peace and solitude within nature. "We want a ground," wrote Olmsted in his most influential essay, "to which people may easily go after their day's work is done, and where they may stroll for an hour, seeing, hearing, and feeling nothing of the bustle and jar of the streets."[4] Regardless of class distinction, New Yorkers could lose themselves in nature while in Central Park.

This idea of a natural refuge amid hectic streets and cramped buildings soon spread. Cities across the country embraced the call for a connection to nature and began to commission the construction of their own parks to satisfy this need. Although it was several decades after Central Park's completion, primarily because of the tumultuous post–yellow fever years of the late nineteenth century that plagued the southern city, Memphis eventually joined the likes of St. Louis, Kansas City, and Chicago in following New York's lead in the urban parks movement. George Kessler, a German-trained landscape architect and protégé of Olmsted's, was selected by the Memphis Park Commission in 1901 to design a series of parks and parkways to integrate the city with its natural surroundings.[5] Designing Overton Park—named after city cofounder John Overton—on the outskirts of Memphis, Kessler crafted a destination that enabled many Memphians to feel as if they had fled the city. Incorporating Olmsted's signature greensward component into the center of the park while also highlighting the beauty of the forest with trails and paths, Kessler harnessed the aesthetic power of nature to enlighten and soothe park visitors.

While the need for a connection to nature initiated and sustained the urban parks movement, many people soon began looking to parks for more than just an escape from the city. Entertainment venues, museums, recreational facilities, and playgrounds quickly filled the open areas that once dominated shared spaces. In Memphis, as in most cities across the country, this focus on increasing the appeal of parks by adding amenities was an attempt to instill civic virtue. By drawing all classes of people together to mingle in the park, whether in a museum or on a greensward, park designers and city leaders felt that they could implement "a system of order and security, an apparently non-coercive means of control and stability." Moreover, they believed that enabling poor and middle-class citizens to rub shoulders with the elite would benefit all of society—a strict paternalistic approach that was typical of early-twentieth-century American leaders.[6] Regardless of the motivation for adding attractions in parks, the physical landscapes of these shared urban spaces began to change as a result of increased amenities. Parks were no longer only a tranquil refuge from the stress of urban life. Instead, they became social destinations.[7]

City parks slowly began to make the transition from the natural landscapes that Kessler and Olmsted envisioned—what the architectural academic Galen Cranz calls "pleasure grounds"—to attraction-filled spaces in the early twentieth century, only a few years after Overton Park had been established. "Park purists," as Cranz calls them, were angered by this change in function and sought to make the alterations as minimal as possible. Describing the attitudes of many early park supporters, Cranz explains that buildings and other structures "were necessary evils to make parks usable. Even their restrained introduction into the parks posed a danger to the pleasure ground ideal."[8] Despite staunch objections from those who wished for parks to exist only as nature sanctuaries for urban residents, progressives pushed on. The pleasure ground of the nineteenth century, although not completely abandoned, became the "reform park" of the early twentieth century.

Unlike their predecessor parks, which focused solely on aesthetics and contact with nature, reform parks placed an emphasis on recreation. Through golf courses, tennis courts, and—perhaps most important—playgrounds, park administrators began offering opportunities for urban citizens to embrace physical activities. While park designers certainly appealed to adults through mediums like golf and tennis, children became the focus of many park initiatives. After 1906, when the Playground Association of America became nationally recognized, playgrounds became commonplace in most urban parks. In many cases, city officials even provided play leaders to facilitate cordial interactions among children on the playground, continuing toward their goal of inspiring civic virtue in park users. One Chicago park administrator went so far as to say that play leaders were some of the most important moral figures a child could have in her life: "Left alone, the tyrannies children practice upon one another compete in cruelty with the oppressions exercised by the brutish governments of grown men," he said. "More ethics and good citizenship can be instilled by a play master in a single week than can be inculcated by Sunday school teachers and Fourth of July orators in a decade."[9] By offering set schemes of physical recreation for park users—restricting children to playgrounds, herding adults onto golf courses, tennis courts, and other fields of play—reform parks contrasted the freedom of pleasure grounds with structure, order, and regulation.

In several ways, Overton Park followed this transition to a reform park. Memphis's first public golf course was built inside the park in 1906, providing a means of "classy" recreation for Memphians not belonging to a country club. This amenity was the first major attraction added to the park, cementing the beginning of Overton Park's gradual transformation. Though

the course was not part of Kessler's original plan, his incorporation of vast open spaces to achieve a connection to nature
made the course's construction simple.[10] Five years after the
golf course was formally established, Overton Park built its
first playground, making a concerted effort to attract children
and families to this space. Like others across the nation, city
leaders in Memphis embraced the concept of play directors.
Offering up well-respected members of the community to
lead play and serve as role models, administrators hoped to
improve the character of countless Memphis children who
used the Overton Park playground.[11]

An early photo of golfers enjoying the course in Overton Park.
Photo courtesy of William Bearden, author of *Images of America:
Overton Park.*

But physical recreation, whether for children or adults, was not the only new component of reform parks. Educational and cultural institutions also sprang up during this era, especially in Overton Park. In 1906, the same year the golf course in Overton Park opened, the Memphis Zoo was officially established. The formal beginnings of the zoo followed a few brief years of an impromptu zoo in the park, with a grizzly bear named "Natch" chained to a tree by Memphis businessman A. B. Carruthers. Allegedly, Carruthers had received Natch as a form of payment for a business transaction. After unsuccessfully donating the bear to Memphis's professional baseball team, the Memphis Turtles, Carruthers was forced to again house Natch on his own property. No longer willing to deal with the demolished flowerbeds that resulted from keeping a bear in his front yard, Carruthers finally "offered" Overton Park as a new home for Natch. From this unwanted gift, a world-class zoo arose, offering educational and recreational opportunities for countless Memphians.[12]

In line with urban parks across the country, Overton Park soon added a cultural institution as well. The Brooks Memorial Art Gallery, built in 1916, offered Memphians the chance to explore their artistic interests.[13] After 1916, park visitors could, theoretically, experience a connection to nature in the forest, play a round on the golf course, educate themselves on different animal species in the zoo, and then enter the Brooks for cultural immersion in paintings and sculpture. Offering multiple formats for recreation, Overton Park—and reform parks across the country—juxtaposed the earlier ideals set forth by park purists. As opposed to simply engaging with nature, Memphians depended upon Overton Park for multiple forms of entertainment and interaction.

Shortly after the Brooks Museum became entrenched as a park institution, urban parks across America began to shift their focus yet again, exiting the reform era and entering an "age of leisure," which began around 1930 and lasted through the mid-1960s.[14] Rather than directing their actions toward goals of civic improvement and moral reform, park administrators and city leaders abandoned attempts to justify park spaces. City dwellers, they realized, now thought of "park facilities as an expected feature of urban life."[15] Especially during the Great Depression and World War II, people relied on parks to provide entertainment and—in many ways—happiness. Through the amenities established during the reform park era, such as museums and playing fields, and through new park attractions, such as open-air concert venues, urban parks helped citizens persevere through difficult times.

Memphians during this era certainly relied on Overton Park in the ways described above. Perhaps the best illustration of this dependence on the park can be seen through the Overton Park Shell, built in 1936 by the Works Progress Administration. Modeled after larger amphitheaters in Chicago and St. Louis, the construction of the Overton Park Shell provided job opportunities for Memphians looking to support their families during the Depression.[16] But its impact was felt long after construction. Hosting light operas, symphony concerts, and dance performances, the Shell—which was also called the Memphis Open Air Theatre, or the MOAT, because of its sophisticated shows—provided free or cheap entertainment to citizens searching for leisure.

In its first four decades, Overton Park underwent significant changes. Transforming from a nature refuge to a social destination, the park gradually became a mixture of Kessler's rural-oriented masterpiece and Progressive Era amenities. Although the park has always been an ever-changing place within the city, its most radical physical transformations occurred during these decades.

A RECREATION HAVEN

Because Overton Park's status as a pure nature haven dates back over a century, no one remembers it in its capacity as a pleasure ground. But several people recall the park as it entered into a new era in the 1940s. Michael Cody, a native Memphian and former Tennessee attorney general, grew up in Memphis as the park was undergoing the transformation from a pleasure ground to a mixed rural-urban destination. He recalls several things that the park offered during his childhood in the 1940s and 1950s. "My first memory was playing in the wading pool in the park, mainly during the summers," he says. The wading pool, as Cody describes it, was a large, concrete oval with a fountain in the center. It was designed to be no more than three feet deep at its deepest point, so even young children could enjoy wading in the water without too much concern from their parents. Torn down in the late 1970s, the wading

Children playing in the wading pool. Photo courtesy of William Bearden, author of *Images of America: Overton Park*.

pool was once located near a popular playground in the park. This area—especially in the scorching summer months—became an exciting destination for children across the city.

In addition to the unstructured running, romping, and wading that occurred in this part of Overton Park, Cody remembers an extensive amount of programming that revolved around the wading pool in the early 1940s. "They had camps and games, and of course," he remembers, "it was all segregated. All these kids were just white kids. They wouldn't allow black children." Cody, who grew up to become heavily involved with the American Civil Liberties Union and the civil rights movement—even representing Martin Luther King Jr. in court the day he was killed in Memphis—was certainly impacted by his childhood experiences with systemic racism in Overton Park.

It would be two more decades before all Memphis children could enjoy playing in the park. Programming during this pre-integration era also included pageants, parades, and plays. Many young girls would dress up in white dresses, he remembers, and participate in the pageants. For those uninterested in joining in the more refined events, adult planners also organized athletic competitions and games. Some of these games were rather unorthodox, but despite their complexity, Cody recalls the joy that these events brought him. "I remember playing 'box hockey,'" he laughs, "but I don't remember the rules."[17]

Though pageants, performances, and "box hockey" certainly had their place in the park, children's recreation was dominated by more popular games. Baseball was especially important to local children. Charlie Newman, who defended the park in the Supreme Court from destruction by Interstate 40, remembers gathering with other neighborhood children for baseball games throughout the spring and summer. Congressman Steve Cohen, a native Memphian, also describes vivid memories of America's pastime in the park: "The Rotary League baseball team with which I was associated played their games at Overton Park. In the '50s, and I guess in the '60s, Rotary League baseball was a big deal." The facilities at Overton Park reflected the seriousness that the players brought to the field. Backstops, small grandstands, plates, bases, and pitchers' mounds adorned the open landscape, remembers Cohen, making the experience all the more meaningful for players and fans. "Baseball was a big deal growing up in Memphis," he reiterates. "It was hardball."[18]

Few will be surprised to learn of baseball's significance to children in Overton Park. After all, this was—and many argue still is—America's game. In the 1940s and '50s, children were stepping up to the plate, playing catch, and stealing bases in city parks all across the country. Golf, on the other hand, held a unique appeal for children in Overton Park during the mid-twentieth century. The Links at Overton Park has hosted a junior golf tournament since the mid-1940s, inviting children from across the region to participate. Steve Cohen and Mike Cody both share fond memories of playing in this youth tournament. Cohen recalls that the best round of golf he ever played came during the preliminary round of the youth tournament. It was both a blessing, he explains, and a curse. "Often, players tried to get a little higher score in their first game so they [could] get into a seed that wasn't quite as strong, and they'd have a better chance of winning their flight. I had the best round I've probably ever had, and I got into a higher flight. So it was good to get that round," he laughs, "but I didn't make it out of the next round." Cody, however, fared better. In 1948, when he was twelve, he won the tournament. "I gave the [Abe Goodman Golf Clubhouse] a picture of me and some of my friends with a golf pro showing us how to hold the club and a clipping of how we all did in that tournament. That clipping," he says with a smile, "is still on the wall in the golf house."[19]

Perhaps the most interesting story of competition and recreation in Overton Park during this midcentury era stems not from first-person experience but from a child watching his father. "When I was young—and I don't know whether I was eight, ten, twelve, whatever my age was," Cody recalls, "they had a big tournament at Rainbow Lake for fly fishermen from all over the South." Strategically placing wooden rings of various colors throughout the lake, the tournament hosts designed a course for the fishermen to navigate with their "popping bugs," special lures used in fly fishing. Standing in one spot, the competitors had to cast from "the yellow one, to the green, to the red, to the blue—they had to follow a pattern. It was just like throwing darts at a target, but they were using a fly rod," Cody explains. Similar to his son's experience in the 1948 Junior Golf Tournament, Cody's father prevailed. "My father won that tournament," remembers Cody. Competing in Overton Park, it seems, was natural for the Cody family.[20]

OVERTON PARK'S AMENITIES: A RISE IN POPULARITY

As evidenced by these rich memories, Overton Park was a favorite spot for recreation and competition. Whether playing baseball or competing in a fly fishing tournament in Rainbow Lake, Overton Park offered ample leisure opportunities. But not all memories from the 1940s and 1950s revolve around recreation. Many stories focus instead on the numerous amenities that the park offered to its patrons. Cody, for example, remembers the Japanese Garden, which was located at the present site of the Memphis College of Art. The Japanese Garden was installed in 1914 and served as a beautiful cultural immersion in the heart of the park. Park patrons could mingle in the gardens and spend an entire afternoon strolling among the lovely sights and scents. After the attack at Pearl Harbor on December 7, 1941, however, the Japanese Garden was vandalized and then officially removed by the city. "They cut all those trees down to the ground because of the Japanese attack," explains Cody. "That was a big thing that I remember. They were beautiful, beautiful trees—quite admired by people all over the region." With anti-Japanese sentiment rapidly spreading across the country, the beauty and peace that the garden offered could not deter its destruction. Many Americans believed that fighting World War II on the home front required eliminating any semblance of Japanese culture from their daily lives.[21]

A January 17, 1942, article from the *Memphis Press-Scimitar* describes the destruction of the Japanese Garden. Photo courtesy of William Bearden, author of *Images of America: Overton Park*.

Around the same time that the Japanese Garden was destroyed, other park amenities began growing in popularity. Chief among these much-loved institutions were the Brooks Memorial Art Gallery, the Memphis Zoo, and the Overton Park Shell. Few people, unfortunately, remember the Brooks during this era. As children, they spent most of their time in other areas of the park, not yet interested in the fine art housed in the museum. Despite the lack of memories, the Brooks Memorial Art Gallery thrived during this period. This success is most evident when examining the expansion of the Brooks's facilities. Originally built in 1916 and dubbed the "Jewel Box of Memphis" because of its splendid architectural design, the museum was filled to capacity by midcentury. To secure a large donation of artistic objects from the Samuel H. Kress Foundation, the museum was forced to expand. In 1955, the Brooks underwent the first of several enlargements. This need for growth stands as a testament to the Brooks's popularity among art enthusiasts from Memphis and around the region.

Unlike the Brooks Memorial Art Gallery, the Memphis Zoo during this period is remembered by an abundance of Memphians. White children, especially those too young to understand the moral and legal injustices of segregation, often felt pure bliss when walking through the zoo, gaping in awe at the animals surrounding them. Highlighting the memories of people who visited the zoo as children in the 1940s and 1950s reveals that this institution has captivated its patrons for decades.

A drawing of the Brooks Memorial Art Gallery before its opening to the public. Image courtesy of William Bearden, author of *Images of America: Overton Park*.

Ernest Kelly was one of those children who walked in amazement through the zoo, often strolling alongside his grandfather. Kelly, a retired attorney and Overton Park enthusiast, remembers that his grandfather specifically used their zoo visits as a bargaining tool when being courted for a new job. When asked to become the leader of an organization that promoted Memphis trade on an international level, Kelly's grandfather hesitated. "'Well, I have a grandson now,'" Kelly recalls him saying, "'and I can't take any job that's going to prevent me from taking him to Overton Park and the zoo every day of the week.'" The bargaining strategy worked. A mornings-only deal was brokered between Kelly's grandfather and his employers, facilitating countless more zoo trips for Kelly.

With daily visits to the Memphis Zoo, Kelly soon became well-acquainted with the lions, elephants, monkeys, and more. "I knew all of the animals by their first name," he claims. Charlie Newman also made daily treks through the Memphis Zoo, though for a different reason. Living in a neighborhood just southwest of the park, Newman walked through the zoo each weekday to get to Snowden School. Newman and Kelly may be unique in the frequency that they visited the zoo. But whether children visited on a daily basis or less often, it is fair to say that the Memphis Zoo attracted scores of children in the 1940s and 1950s.[22]

On occasion, the Memphis Zoo would extend beyond its borders. It was normal, explains Kelly, to hear the lions roaring from his home each night. Peacocks, too, were frequently heard throughout the surrounding neighborhoods. More exciting—and without a doubt more frightening—was when animals would escape from their exhibits. Kelly recalls once seeing what he thought was a pack of large dogs walking around his home: "We later found out that they were black timber wolves." Michael Cody remembers a slightly more comical escape experience from the late 1950s. While a student at Southwestern, which is now Rhodes College, Cody says that some fraternity members—though, he is quick to point out, not members of his own fraternity—snuck into the zoo with long, wooden planks. Leaving the college that was just across the street and heading straight for Monkey Island, the students extended the planks over the fence and moat that surrounded the island, creating an escape path for the monkeys. "The next morning," laughs Cody, "there were monkeys all over the zoo." For the most part, though, Memphians enjoyed the zoo from inside the facility. Its popularity during this era helped attract countless park visitors and paved the way for continued growth and success.[23]

Despite the pleasant stories of walking to the zoo with family members or animals escaping to nearby backyards, it is crucial to note a defining characteristic of the Memphis Zoo in the 1940s and 1950s. Like all park institutions, from the Brooks and the Shell to the wading pool and playground, the Memphis Zoo was a segregated space. Only white Memphians could enjoy the park and its institutions whenever they pleased.

Johnnie Turner, a Tennessee state representative and outspoken civil rights activist, remembers when she and her family were only allowed to attend the zoo on one day of the week. "[If you were black]," she explains, "you could only go

Johnnie Turner, top right, visiting the zoo in 1959 with
family members before this amenity was integrated.
Photo courtesy of Johnnie Turner.

to the zoo on Thursdays." The signage at the zoo was a blatant reminder that black and white people were not supposed to intermingle. On Thursdays, says Fred Davis, another long-time civil rights proponent and a contemporary of Dr. Martin Luther King Jr., there was a sign that went up at the zoo

The prominent African American photographer Ernest Withers captured this image of a sign that helped enforce racial segregation at the Memphis Zoo. © Dr. Ernest C. Withers Sr., courtesy of the WITHERS FAMILY TRUST, thewitherscollection.com.

that read, "No Whites Allowed." Whereas signage during the pre-desegregation era typically barred blacks from entering white establishments, the city-funded zoo wanted to ensure that black citizens were not given the opportunity to visit alongside whites on the one day they were allowed to enter. By restricting access for whites on Thursdays, the zoo was not infringing upon the rights of white people—rather, they were making a concerted effort to safeguard racial segregation.[24]

In the face of constant reminders of segregation, Turner still found a way to enjoy her trips to the Memphis Zoo. "We loved coming to the zoo. Even though it was segregated, it didn't bother me because nobody was there other than other African Americans. That was the usual." The only part that bothered her was the bus ride to the zoo: when she had to move to the back of the bus to give a white person her seat, she was reminded that others felt superior to her because of their skin color. Like other black people across the South, Turner's experiences with public transportation shaped her life and served as yet another motivation to lead the fight for equal treatment in the 1960s.

Once Turner got to the zoo, she felt better because she could just "watch the animals and enjoy life." Like countless other Memphians, Turner went with her family to look at the wide variety of animals. "That was our family outing for me," she says. "I was the oldest [sibling], and every year, I would volunteer to take my brothers and sisters to the zoo." From these trips to the zoo, Turner gained some of her most prized possessions. Because her family had little money, she explains, they could not afford a camera of their own. But at the zoo, staff took photos of visitors with a zoo-themed background and offered the images to families. "These pictures [from the zoo] are treasures to me," she beams. "My photos are priceless because these are the only pictures I have of us having a good time."[25]

It is tragic that the zoo, like all park amenities, discriminated on the basis of race. All throughout the city, whether in Overton Park or elsewhere, the story of segregation remains a wound that is difficult to heal. In both its past and present, Memphis still wrestles with the painful implications of racism and segregation.

The Overton Park Shell, like the Memphis Zoo, was incredibly successful around the mid-twentieth century. It was built in 1936 by the Works Progress Administration, part of President Franklin Roosevelt's New Deal, which sought to revive the failing economy and provide employment for struggling Americans. The Shell quickly became a hallmark within Overton Park. Beginning with more sophisticated performances, such as light operas and symphony concerts, and then hosting a variety of popular concerts, the Shell offered a vast amount of entertainment. During the 1950s, the Overton Park Shell even helped spark the careers of two local Memphis musicians: Elvis Presley and Johnny Cash.

Elvis Presley flashing a smile backstage at the Overton
Park Shell. Photo courtesy of William Bearden, author
of *Images of America: Overton Park*.

Elvis's rise to stardom is intimately tied to the Shell. On
July 30, 1954, the young singer took the stage to open for Slim
Whitman. It was the first paid performance of his legendary
career. A hometown hero, Elvis appropriately cut his teeth
on a hot summer night on a concrete stage in the middle of

Memphis. The Overton Park Shell was a catalyst for Johnny Cash, too. "Elvis Presley asked me to sing with him at the Overton Park Shell in Memphis, and I sang 'Cry, Cry, Cry' and 'Hey Porter,'" Cash said in a 1997 interview with National Public Radio. "And from that time on, I was on my way, and I knew it, I felt it, and I loved it."[26]

In addition to stars such as Johnny Cash and Elvis Presley, many other singers, actors, and dancers performed at the Overton Park Shell. Blanchard Tual, a Memphis lawyer who served as president of the Shell from 2012 to 2013, remembers watching his father in performances that were part of the Memphis Open Air Theatre (MOAT) productions. "I can remember going and seeing him on stage," Tual recollects. "I have vivid memories of that." Though he cannot remember the names of the productions his father participated in, it is likely that his father performed in one of many light operas, which were among the most popular genres performed at the Shell during this era. The light opera, explains Charlie Newman, was "a cross between opera and Broadway musicals. They always had a lot of that at the Shell." Newman remembers hearing the beautiful sounds from these performances from his home near Overton Park.[27]

Blanchard Tual's experience of watching his father perform at the Shell is unique; most Memphians sat with their family members in the audience instead of watching them on stage. Steve Cohen is one of many who recollects attending MOAT events with his family. "I went to [performances] with

A 1948 advertisement for the Memphis Open Air Theatre.
Photo courtesy of William Bearden, author of *Images of
America: Overton Park*.

my parents. I have early, early memories of going to concerts there," he says. Ernest Kelly has perhaps the most compelling memories—some happy, some sad—of attending productions at the Shell with his family. He remembers a particularly rainy evening when, as a young child, he accompanied his family to a light opera performance at the outside venue: "We were the last holdouts—it was pouring down rain at the MOAT. It became pretty obvious [that they were going to cancel], and almost everybody else was gone. We were still waiting there, hoping against hope that the rain would stop and the play would come on." Wearing his grandfather's hat helped keep him dry as they sat waiting. The show, unfortunately, did not go on.[28]

Another of Kelly's memories involving the park and his grandfather is much less lighthearted. Kelly's grandfather was, along with his father, his first hero. The friendship they had built was strong and lasting. And the MOAT performances at the Shell, and Overton Park in general, were integral to nourishing this friendship. They attended concerts together summer after summer, always walking through the neighborhood and then through the park before finding a seat in the audience. The night before Kelly left for his senior year of college, he attended his last Shell performance with his grandfather. With the exception of a morning send-off at the airport, it was the last time he saw his hero. "He was, I guess, eighty-eight at the time. He paused to take an angina

Ernest Kelly, right, wearing his grandfather's hat as he
and his family tried to wait out the rain at the Shell.
Photo courtesy of Ernest Kelly.

and nitroglycerin pill as we came up the hill, and [then we]
walked over to the last Starlight Symphony that we went to
together. Later that fall, I got the call that he had died." For
Ernest Kelly—and for many other Memphians—the Shell
will always be a source of memories.[29]

CONCLUSION

As evidenced by the many memories of this public space from the 1940s through the 1950s, Overton Park was a place of increasing importance. The park was where people went for entertainment, for recreation, and for relaxation. They also went to the park for connection. Whether watching a parent compete in a fly-fishing tournament or walking to watch a light opera at the Shell with a grandparent, Memphians connected with both the park and each other. The stories of these individuals are inseparable from the story of the park itself.

As the 1950s came to a close and the 1960s began, the park entered into a period of controversy. A bitter fight between the highway department and park users heated up near the end of this decade, which was full of demonstrations, lawsuits, and death threats. Interstate 40 threatened to destroy the park that many Memphians had come to love. But the controversies cannot be limited to the beginnings of a famous legal and environmental struggle. Larger American issues, such as the antiwar movement and the civil rights movement, also consumed Overton Park. Most important was the fight for racial equality. As they did in cities across the nation, black Memphians increasingly began demanding fair treatment. The civil rights movement was a turning point in Memphis history, and Overton Park was an important site for that fight for equality. The Overton Park of the 1940s and 1950s was about to change.

CHAPTER 2

CHARACTERIZED BY CONTROVERSY
The 1960s

The Overton Park of the 1940s and 1950s is often remembered as a simple space that provided endless leisure and
recreation and facilitated joy among both children and adults.
Stories revolve around trips to the Zoo, walks through the
wading pool, and awe-inspiring performances at the Overton
Park Shell. Segregation was not yet widely challenged during this era, and the park—with the exception of the Japanese Garden's destruction in the wake of the attack on Pearl
Harbor—seemed to be free from political and social issues.

In 1960, Overton Park entered into a decade defined by
controversy. The park became a key component in several
widespread movements, providing citizen-activists a setting

in which to push for their causes and beliefs. Overton Park, for example, was carefully utilized by black Memphians who looked to end segregation in the city. Because the park was a public facility funded through taxes paid by all Memphians, black leaders reasoned that it should be an early target of desegregation attempts. After all, what is a public institution that is not open to the entire public? Rather than focusing on integrating private businesses, as activists did at lunch counters in Greensboro and Nashville, black leaders in Memphis put pressure directly on the city government by concentrating on public areas. In addition to civil rights demonstrations, the park later became a popular site for protests against the widely unpopular war in Vietnam. To promote their cause, Memphians made speeches in front of hundreds of attendees, calling for peace overseas. Throughout the 1960s, citizens used this space for a variety of political causes, always cognizant of their location within the park and the city.

The final major political controversy of this decade, though only in its early phase, was an attack on the park itself. With Interstate 40 inching ever closer to destroying the Old Forest in Overton Park, citizens mobilized to protect the land they loved. Calling themselves the Citizens to Preserve Overton Park, or CPOP, these Memphians organized in 1957 and promised to continue fighting the highway department's decision to bulldoze swaths of forest to install an expressway. For about ten years, the debate ebbed and flowed, gaining momentum at points and losing it at others. Near the end of the 1960s, the dispute with highway officials intensified to new levels. Although the now famous Supreme Court case would not occur until 1971, CPOP was already in a skirmish to defend Overton Park. In the 1970s, this skirmish would become a full-fledged war.

Out of these multiple controversies, Overton Park started slowly growing into Memphis's park. After successful sit-ins that coincided with victories in the courts, black Memphians gained the right to use the park and its facilities whenever they pleased. Because of heroic efforts by many ordinary citizens, the park was no longer a segregated space. Antiwar protests helped create ownership of the park, too, solidifying Overton Park as a place for Memphians to voice their opinions to the public. Even the interstate crisis proved to be beneficial to the park in the long run. By forcing citizens to band together in the face of almost insurmountable adversity, the proposed expressway fostered in many Memphians a sense of responsibility to be good stewards of Overton Park.

THE CIVIL RIGHTS MOVEMENT IN OVERTON PARK

Throughout the United States—and especially in the South— African Americans amplified their efforts to fight discrimination and segregation in the early 1960s. Only two generations removed from ancestors who had been forced to endure a life of slavery and servitude, black Americans were still treated as second-class citizens in the mid-twentieth century. Countless efforts had been made by African American leaders prior to the start of the civil rights movement to secure equal treatment. Activists such as Frederick Douglass, W. E. B. Du Bois, Ida B. Wells, James Farmer, and Amzie Moore, among many others, pushed for an increase in rights and battled discrimination on personal and systemic levels. But sadly these attempts yielded few tangible results. During the 1950s, this battle for equality became more organized and visible. Rosa Parks inspired others to take action when she refused to give up her seat on a bus in Montgomery, Alabama, in 1955, and Ella Baker, John Lewis, Medgar Evers, and Martin Luther King Jr., among others, helped organize nonviolent protests with newly motivated citizens to demand full rights for all Americans regardless of race. By the early 1960s, the battle against segregation and discrimination had evolved into a successful and revolutionary movement.

Memphis was an important site of the civil rights movement, not just because it is the city where Dr. King was assassinated. Civil rights activists in Memphis attempted to desegregate all public buildings and events in the early 1960s. Targeting these places, says Fred Davis, a lifelong civil rights leader in Memphis, was part of a deliberate plan: "It's illegal to use all citizens' taxes to support an institution that other citizens only have limited use of." Davis invokes this legal and political injustice to explain why he and fellow activists orchestrated their desegregation strategy as they did.[1]

Johnnie Turner, who was also actively involved in public facility sit-ins, affirms Davis's reasoning. "When I bought a Coke, the tax that I paid on that was being used to subsidize the Brooks Gallery. It was being used to subsidize the Overton Park Shell, which is now Levitt Shell. It was most certainly being used to subsidize the zoo," she says. "It was public money serving [these institutions]." As a dynamic public place in the heart of the city, Overton Park and its amenities offered activists in Memphis an obvious target for integration. Seizing upon the ideal moment, a small group of black students made their move, bringing the civil rights movement into Overton Park.[2]

Johnnie Turner was one of the students determined to effect change. Turner's passion for racial justice can be traced all the way back to her experiences as a child: "I have so many memories of what life was like when segregation prevailed— when I was made to feel less than human." When she was only four years old, Turner and her family fled to Memphis from a brutal sharecropping plantation in Arkansas. To escape the cruel mistreatment inflicted upon them—mistreatment that left the Turners only one rung above slavery—they had to climb through barbed wire fences and swim across a lake under the cover of midnight. Until death, explains Turner, her mother carried a deep scar on her leg from climbing through those fences. The scar, though painful, served as an inerasable reminder of a "successful escape from servitude." Turner's experiences as a young girl, along with her encounters with racism on buses and in schools once she arrived in Memphis, instilled in her an abiding desire to fight oppression, inequality, and discrimination.

When Turner enrolled at LeMoyne College (now LeMoyne-Owen College), a historically black institution in Memphis, she quickly became involved with the sit-in movement. "We started out in the [public] libraries. . . . We figured we had a better chance for having success if we started out in that vein," she says, reiterating the strategy that local civil rights leaders had developed. Even though she and her classmates were excited to be involved, Turner still felt anxious about being an active participant in these sit-ins. "I didn't even tell my daddy," she says, worried that her father would think she was jeopardizing her full-ride scholarship to LeMoyne. But it was through participating in these demonstrations across the city—demonstrations that also targeted the Brooks Museum in the park—that Turner met the other students who would join her in the 1960 sit-in, or "kneel-in," at the Overton Park Shell.[3]

On August 30, Turner and other young activists attempted to integrate a Youth for Christ rally at the Shell—the same venue where Elvis Presley had, only six years earlier, blended black and white music together to bring rock and roll to the world. According to Turner, the advertisement for this religious service read, "Youth for Christ rally—Open to the Public." Determined to see whether they would be allowed to worship at this event, given the broad scope of who was invited, the students resolved to attend the service. Turner remembers thinking, "They won't arrest us—that's a church!" Steadfast in their faith, confident in their mission, and committed to a Christianity that purported to transcend boundaries of race, Turner and the others put on their best clothes and headed to the Overton Park Shell.

When they arrived, explains Turner, they were not initially allowed inside. The student group's leader, Evander Ford, explained their intention to an usher at the rally. As Turner remembers, "Ford said, 'We just came for the service. [It's a] Youth for Christ rally, and we're youth. We just want to hear the service. We're not coming to disrupt anything.'" After this brief explanation, the students were allowed to enter. Turner recalls that, at the beginning, everything went as planned. "I went in and sat down, and I was really enjoying [the service]—I really was." The collection plate was passed around, and Turner put in $1, which was, she explains, a lot of money for her at the time. Despite the cold welcome they initially received, it seemed that the students would be allowed to stay and worship as they had hoped.

But that optimism crumbled almost immediately after Turner had given her offering. "I got a tap on the shoulder," she remembers, "and this police officer said, 'Come with me.' I was so shocked." She and the other students were escorted out of the Shell and put in the backseat of a police car. Although she was quiet and respectful during the entire time she was in the Shell, she was being arrested for "disturbing" a religious service. But unlike the others, Turner was held overnight in jail. "They told me they kept me because I was a criminal. And they interrogated me to find out why I [participated in sit-ins]," she says, citing her previous record of arrests for involvement in civil rights demonstrations as the main reason she was forced to stay overnight. "I was just petrified. I was scared to death. But I never said a word." Through the initial arrest, interrogations, and overnight stay, Turner was still in her church clothes. "I slept all night in my beautiful white dress."

In spite of the fear Turner felt after her overnight stay in jail, she pushed forward and refused to let this experience stifle her passion for pursuing civil rights. If anything, her experience in the Overton Park Shell motivated her to be even more vocal and proactive in the movement: "I was at the 1963 March on Washington. I was there when Dr. King gave his brotherhood masterpiece, 'I Have a Dream.'" And five years later, her involvement was just as strong: "I was at every march Dr. King led in Memphis. I was right there. When Dr. King gave his prophetic 'I've Been to the Mountaintop' speech, I was sitting right there."[4]

Johnnie Turner, far right, smiling with her family at the
Memphis Zoo in 1967 after the space was integrated.
Photo courtesy of Johnnie Turner.

Turner's unrelenting courage and ceaseless fight for racial
justice helped bring change to Memphis. Her impact was
especially apparent in Overton Park. Three years after she and
her classmates attended the Youth for Christ rally at the Overton Park Shell, black Memphians achieved a monumental

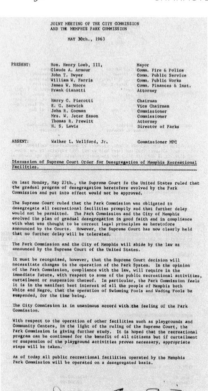

A letter from the Memphis Park Commission regarding the Supreme Court's decision in *Watson v. City of Memphis*. Park Commission members decided to suspend the operation of some public facilities, like swimming pools and wading pools, rather than operate them as integrated spaces. They also considered closing playgrounds instead of immediately integrating these public areas. The letter demonstrates the perverse reluctance of some white city leaders toward racial integration. Photo courtesy of the Memphis and Shelby County Room, Memphis Public Library & Information Center.

victory in the courts. In *Watson v. City of Memphis* (1963), in which Justice Arthur Goldberg wrote in the majority opinion that the court recognized "an unmistakable and pervasive pattern of local segregation" in Memphis, the Supreme Court concluded that all public facilities in Memphis must immediately integrate: Overton Park could no longer legally bar black citizens from entry.[5]

Yet, just because the park was legally integrated did not mean that African Americans immediately flooded the park. Though the law changed swiftly, the culture of acceptance was a more gradual process. Social and cultural norms, explains Turner, "don't just change overnight." Nevertheless, the brave students who helped desegregate the Shell and the park— along with countless other advocates of racial justice—made Memphis an integral city in the civil rights movement, long before King's assassination in 1968.[6]

AGAINST THE WAR IN VIETNAM

Like the civil rights movement, opposition to the Vietnam War stretched far beyond Memphis and Overton Park. Distraught that the United States had invaded Vietnam for what they saw as an unjust cause, many American citizens were quick to voice their disapproval, especially as the war dragged on. In demonstrations across the country, protestors denounced the United States' involvement in the war by burning their draft cards and railing against the government for what they believed was a horrendous moral and political transgression. To this day, the war in Vietnam remains one of the most controversial and unpopular political decisions in American history.

Overton Park was an important location in Memphis for antiwar activists to voice their concerns and advance their protest. Michael Cody, who was a young lawyer and a national board member of the American Civil Liberties Union in the late 1960s, was an essential part of the largest antiwar event that the park hosted. Cody explains that he consistently worked with groups and individuals who were both strong supporters of civil rights for African Americans and strong opponents of the Vietnam War. As a lawyer, he even provided counsel to young men who wished to avoid the draft. Because of his involvement in the antiwar movement, he was an ideal candidate to give a calm, measured, and convincing anti–Vietnam War speech in front of a large crowd in Overton Park. In 1969, he did just that.

In 1961, Cody himself had actually volunteered to join the military. "[I knew that] when I got ready to get out of law school, I would be drafted," he explains. "So that's when I joined the army and served six months' active duty and then five and a half years of reserve. I still would have been drafted [after law school] if I hadn't joined the army," he says, but volunteering when he did kept him from serving on the front lines in Vietnam. While enlisted from 1961 to 1966, he worked in army intelligence during the Cold War, when the United States' primary enemy was the Soviet Union. "My job was to interrogate Russian prisoners of war. Of course," he laughs, "we never went to war with Russia, and there weren't any [prisoners of war]. So I drove a jeep around and played soccer." Despite Cody's limited involvement, his military service gave him a unique perspective on the Vietnam War, a perspective that was valuable for his speech in Overton Park.

Speaking to a large group of citizens with Rainbow Lake as his backdrop—the same lake where his father had once won a fly-fishing tournament in the 1940s—Cody delivered what was the only antiwar speech he ever made. He expressed his disapproval of the Vietnam War and drew on personal experiences to inform his rhetoric. His familiarity with conscription played an important role in his distaste for the war. Though he avoided the draft by volunteering for the military before the war intensified, he was involved closely with others who felt firsthand the impact of conscription. Even beyond his role of providing legal counsel to several young men who hoped to avoid the war, Cody had many friends who were drafted. "In those days," he explains, "if you weren't in school, you got drafted. If you got drafted, you went to Vietnam. A lot of people went to Vietnam who didn't believe in the war." As he spoke at the gathering in Overton Park, Cody's own unique experiences with the draft and the military added layers of authenticity, credibility, and persuasiveness to his speech. Though like-minded activists found inspiration in his address, other citizens responded in a less accepting fashion. "There was a big reaction," Cody remembers. "Most people living in Memphis thought that was pretty close to being a communist in that period of time."

Michael Cody speaking to a large crowd of antiwar activists. The Greensward is in the background, and Rainbow Lake is just out of the picture to the right. Photo courtesy of William Bearden, author of *Images of America: Overton Park.*

As with all controversial political issues, Cody took a risk by publicly voicing his beliefs. But by standing up for what he believed in, he inadvertently helped establish Overton Park as a setting for citizen activism. Today, Memphians still use this space to express their ideas in a public location, whether to support the Black Lives Matter movement or to oppose parking on the Greensward. Like Turner's efforts in the civil rights movement, Cody's courage to push back against what he felt was an unjust cause in the late 1960s undoubtedly shaped the way that Overton Park was utilized by activists in the decades to come.[7]

THE EXPRESSWAY BATTLE INTENSIFIES

Civil rights and antiwar activism had substantial impacts on Overton Park's patrons, redefining who could use the park and how they could use it. The expressway controversy, on the other hand, directly affected the parkland itself. In the mid-twentieth century, a highway boom began in the United States. Funded mostly by the federal government, these new roadways were intended to connect major cities across the country, both to benefit millions of civilians traveling long distances and to strengthen military infrastructure during the Cold War. Because parkland was already publicly owned—and would thus reduce the government's cost for constructing highways—it became a popular target for transportation officials who preferred to use what was convenient and cost-effective rather than purchasing private land. This was certainly the case in Memphis. In 1956, the Federal Bureau of Public Roads approved a plan to route Interstate 40 directly through Overton Park. This plan, if enacted, would connect East Memphis with downtown and provide a direct path through the city for both the trucking industry and citizens. But it would also destroy a sweeping expanse of the Old Forest, forever changing the character of Overton Park.

Although many Memphians were angry that the interstate was slated to be built through Overton Park, a small group of determined citizens channeled this anger into action. Organizing in 1957 and calling themselves the Citizens to Preserve Overton Park (CPOP), a handful of activists began to protest the proposed interstate route. Sensitive to the public outcry that came from the proposal to run the interstate through a beloved public space, the highway department worked on completing other sections of the interstate. But in 1964, just one year after the park was integrated as a result of courageous activism and the Supreme Court decision in *Watson v. City of Memphis,* highway officials again keyed in on Overton Park. To their credit, they proposed a new plan that would lessen the interstate's impact on the park, erasing only twenty-six acres of the park rather than several hundred. However, their plan would still cause significant damage to the park, which led CPOP to refuse to submit to any portion of this space being transformed into a paved roadway. The lines were drawn, and the showdown between the government and CPOP escalated.[8]

Charlie Newman, who served as CPOP's attorney when the legal battle began, remembers clearly the early stages of the controversy. The year 1966 proved to be important, he recalls, especially in regard to the newly created legislation on which CPOP later based their critique of the interstate. In October, Congress passed the Department of Transportation Act of 1966. One section of this vast legislation proved essential to CPOP's cause: Section 4(f). Newman explains this section concisely: "Basically, federal money could not be used to build a federal highway . . . on parkland or certain other kinds of land, like historically significant land, unless there was no feasible and prudent alternative." And if no "feasible and prudent alternative" existed, then the highway officials must "use all possible planning to minimize harm to the park." With the passage of this legislation in 1966, CPOP gained legal ammunition to continue their fight.

For a few years, CPOP continued to protest the interstate as they had for the last decade: they attended public meetings, printed bright yellow T-shirts to raise awareness about their cause, and spoke with anyone who would listen in an attempt to gain more support for the park. In 1969, they shifted their strategy. Newman recalls that less than a year after he, Michael Cody, Lucius Burch, and David Caywood had represented Dr. Martin Luther King Jr. in court the day King was assassinated, he received a call from a friend who was a lawyer in Washington, D.C. "[My friend's] partner had been hired by the Citizens to Preserve Overton Park. This guy's name was Jack Vardaman, and he had been hired in Washington because [CPOP] had gotten in touch with a national civil rights organization," Newman explains. The Citizens to

Preserve Overton Park had filed suit against the secretary of transportation, John Volpe, in Washington District Court on December 3, 1969. Though the case would be transferred to Memphis in early 1970 because the Tennessee Department of Transportation was added as a defendant—it was at this point that Newman became involved at Vardaman's request—the beginning of the legal battle in Washington remains crucial to CPOP's story. No longer willing to settle for small-scale activism, the Citizens to Preserve Overton Park had legally challenged both the federal and state governments.

All during this decade, both before and after CPOP had filed suit, Overton Park's defenders had met strong resistance from select groups. Though city government was initially in favor of finding a route for Interstate 40 that avoided Overton Park, that vision soon changed. "At some point prior to 1969," Newman recalls, "the federal highway people and the state highway people had persuaded the Memphis City Council that if they didn't give in to the [Overton Park] proposal, they would lose all of their proposed funding and would not get a highway. So the City Council reversed itself and unanimously supported the highway."

But federal, state, and local governments were not the only opposition that CPOP encountered—the trucking and transportation industry also constituted a significant opponent. Acting as "fierce, fierce lobbyists," Newman says, transportation companies fought hard for the interstate to be constructed through Overton Park. The incentive was economic: the longer the interstate's completion was delayed, the more money these industries lost. The tactics used by these transportation-based industries became more aggressive once the legal battle entered the 1970s and CPOP gained momentum. Yet even in CPOP's early stages of protest, the citizen group faced off against a powerful industry that was hell-bent on running an expressway through Overton Park. These Memphians nurtured persistence and determination for over a decade in an attempt to save the park.

In a somber tone, Charlie Newman explains that most of the people who led the Citizens to Preserve Overton Park in the 1960s have passed away: "Mrs. Anona Stoner and her husband, Dr. Arlo Smith and his wife, the Deuprees—they have all died.... They deserve a lot of credit for what they did." Months after Newman's interview for this book, another key CPOP member, Sunshine Snyder, also passed. These activists' insight into the controversy would have offered compelling memories of what it was like to protect the park from the expressway in the early stages of the dispute. Even without their testimonies, though, the story of CPOP's evolution—from designing yellow T-shirts to legally confronting the government—is compelling. When the legal challenge began, Newman remembers, "we didn't have great optimism that we'd be able to stop the highway." But the 1970s offered a continuation—and conclusion—to the conflict.[9]

CONCLUSION

Political and social happenings in the park were of paramount importance during the 1960s. But not every moment in Overton Park during this decade can be characterized by activism and political controversy. The park, after all, remained a *park* throughout these years. While some citizens used this space to fight for racial equality or express a political stance, others ran through the forest, visited the Brooks Museum, and attended the Academy of Art. Children also continued to swarm the park, embracing the outdoors and all that the park had to offer. Gigi Wischmeyer, a longtime park user, remembers playing on the playground as a young girl. She most clearly recalls the metal slide that was located there. It was extremely tall, she says, and "it burned the back of your legs because it was always set right in the sun." Jimmy Ogle affirms the slide's height and explains that the swings on the playground were unusually high as well. Ogle, who spent several years working for the Memphis Park Commission as a special events supervisor, now understands this long-removed daredevil play equipment from a parent's perspective: though many of the kids never hesitated to slide and soar to new heights, watching them must have been "terrifying for the adults."[10]

Sally Jones Heinz, the current executive director at the Metropolitan Inter-Faith Association in Memphis, also remembers playing in the park, although she and her friends

Children braving the famed high metal slides in the
playground. Photo courtesy of William Bearden,
author of *Images of America: Overton Park.*

used their imaginations to create their own playground equipment. Riding their bikes over, Heinz and her friends would
lounge around in the park and then play in some of the many
monuments. The monuments in the park, which range from
honoring Jenny M. Higbee, a beloved educator in Memphis,

to Edward Hull Crump, the notorious political boss of the city in the early twentieth century, were mysterious to Heinz as a child. "[We thought] they were those funny things that you didn't quite know what they were there for, but we would find reasons and ways to play in them," she says. Heinz's, Ogle's, and Wishmeyer's memories of Overton Park during the 1960s suggest that this space continued to be a place for recreation and leisure.[11]

Despite these memories of the park as a place for entertaining children, the 1960s in Overton Park was a decade defined

A statue of notorious political boss E. H. Crump near one of the park's entrances. Photo courtesy of William Bearden, author of *Images of America: Overton Park.*

by activism. The park was, to put it simply, transformed during this era. As a result of their bravery and determination, black Memphians secured the right to use the park without restriction, and they began to feel a sense of ownership of this place, as evidenced by a prominent black newspaper's call to protect the park from destruction.[12] The purpose of the park was transformed as well. No longer was it only a location for recreation and leisure—it was now also a space for public protest and proclamation, no matter how controversial the ideas were. And through the civil rights movement and the interstate controversy, park users during the 1960s effectively became park *owners.* By taking responsibility for the park's wellbeing, the citizen-stewards who opposed segregation and the interstate preserved the park for posterity. Through multiple controversies and confrontations, Overton Park began the slow process of morphing into Memphis's park. But stewardship of the park by its users continued to grow in importance through a series of momentous challenges in the 1970s.

CHAPTER 3

DESERTED BY GOVERNMENT, SAVED BY CITIZENS

The 1970s

I n the 1960s, local activists pushed Overton Park to trans-
form. By opening the park to all Memphians regardless of
race, they altered the composition of visitors and created a
more diverse and inclusive space, even if this transition lacked
haste and enthusiasm on the part of some whites. Through
protests and public gatherings, Memphians also reshaped the
aims of Overton Park. After the 1960s, the space became a
public forum in addition to a recreation, leisure, and nature
destination. Citizens to Preserve Overton Park sparked a
transformation in Overton Park during this decade as well.
By stepping in to defend the park from Interstate 40, this
small group of determined citizens began to foster a sense

of park ownership among Memphians. More so than ever before, park users began to feel that Overton Park was *their* park, and they treated it as such. This revolution in how the park was viewed cannot be underestimated.

Change, then, was a critical element of Overton Park in the 1960s. But once the 1970s began, transformation was the last thing that park patrons wanted. In this era, "transformation" meant that the park would go from being a vibrant, peaceful location within the heart of Memphis to becoming a bustling area with cars and trucks hurtling by on an expressway. Standing against determined government bureaucracies, one-sided media outlets, an aggressive transportation industry, and even fellow Memphians, Citizens to Preserve Overton Park resisted the takeover of their public space. With craft and determination, this group took their fight to the nation's highest court, and after securing a victory in *Citizens to Preserve Overton Park v. Volpe* (1971), they persisted in their efforts to save the park for another decade. Their resolve and passion enabled them to protect the park they loved and to act as the stewards they felt they could—and should—be.

In the wake of cpop's Supreme Court victory in 1971, Overton Park boomed in popularity. Shell concerts rocked—or, in some cases, soothed—audiences. Children continued to swarm the playgrounds, and runners trekked through the Old Forest without the constant roar of interstate traffic. The place grew in vibrancy and was embraced by citizens from across the city. At the same time, however, the park itself began to deteriorate, as did its image among Memphians. Some citizens who were involved with the park during this era think that because the city government was convinced that a highway would be built through the park, the space was purposely ignored. Why would they care for a park that was destined to be covered in asphalt? As a result, the park in the 1970s experienced a simultaneous rise and fall. Defeating the expressway and embracing the park more than ever before represented a peak in the park's history, yet the park slipped into a downward trend that would last for a generation. The years from 1970 to 1980 were filled with ups and downs as the park and its users tried to adjust to life after *Citizens to Preserve Overton Park v. Volpe.*

THE SUPREME COURT AND BEYOND: A DECADE OF ARGUING AGAINST ASPHALT

By 1970, Citizens to Preserve Overton Park had been fighting the construction of Interstate 40 through the Old Forest for over a decade. In the beginning, their activism remained largely localized. They printed flyers, attended public meetings, and wore their signature T-shirts across town. In December 1969, their tactics changed. Challenging the U.S. secretary of transportation, John Volpe, in district court in Washington, D.C., CPOP partnered with attorneys to use the law to aid their cause. Soon after they filed suit, the Tennessee Department of Transportation was added as a defendant, and the case was transferred to district court in Memphis. At this point, the legal contest truly began.

Charlie Newman, who was a young attorney when he was asked to represent Citizens to Preserve Overton Park, recalls that this legal challenge began without much optimism for defeating the highway officials who wished to route the expressway through the park. "When we started the case, we had no reason to believe—and didn't believe—that our chances of success were very high." After all, CPOP and its attorneys were "operating initially on the basis of one law." This one law, Section 4(f) of the Department of Transportation Act of 1966, stated that the United States government must make special efforts to preserve and protect "the natural beauty of the countryside and public park and recreation lands." The secretary of transportation, the statute continued, could only approve projects that would damage these landscapes if no "feasible and prudent" options were available. And if no such options existed, then the project "must include all possible planning to minimize harm to the park." Newman's task, then, was threefold: first, to show that reasonable alternative routes did indeed exist; second, to prove that highway officials simply ignored these other options; and third, to argue that, if the interstate *was* to be built through the park, it had to be made as unobtrusive as possible.[1]

Newman's and CPOP's pessimistic intuitions at first proved true. When they presented their argument to a Memphis judge, they did not receive their desired outcome. "As we expected, the other side [federal highway officials and the Tennessee Department of Transportation] made a motion to dismiss for summary judgment," says Newman, "and the judge granted it, saying that the U.S. secretary of transportation had broad discretion under the law to decide whether or not [the Overton Park route] was appropriate and that the court would not intervene."[2] Essentially enabling the secretary to interpret this statute however he pleased, the Memphis district court dealt a blow against CPOP.

Newman and CPOP did not quit after this initial setback. "We appealed the decision to the court of appeals in Cincinnati, the Sixth Circuit Court," notes Newman. "That's a three-judge court, and they confirmed the decision two to one." Though CPOP lost again, the one vote in favor of their side was crucial. "We got one dissenting opinion from Justice Anthony Celebrezze, who read the statute as having more teeth than that." Finally, a judge had interpreted Section 4(f) narrowly, believing that the secretary of transportation must devote more time and attention to finding an alternative route that would not damage the park. This lone dissent gave Citizens to Preserve Overton Park a small amount of momentum as they headed to Washington to argue their case before the United States Supreme Court.

Aerial view of the construction of Interstate 40. The interstate was intended to slice through, and essentially destroy, Overton Park. Photo courtesy of William Bearden, author of *Images of America: Overton Park*.

Just as Justice Celebrezze had done in the court of appeals, the Supreme Court sided with Citizens to Preserve Overton Park. "The Supreme Court decision said that you cannot excuse or discount an alternative on the grounds that it's not feasible, unless it's virtually impossible to build as a matter of engineering. If you had the Pacific Ocean on one side and a mountain range on the other," jokes Newman, "then perhaps there would be no way technologically to do it. But [their decision] pretty well removed infeasibility as an excuse." With infeasibility dismissed, the Supreme Court moved on to determine whether other alternatives were "prudent." The court defined this term in a way that benefitted Citizens to Preserve Overton Park. Paraphrasing Justice Thurgood Marshall's majority opinion, Newman says the court believed that "Congress would not have passed the law in question if it didn't mean that prudent meant something other than just less expensive and less disruptive. It would always be more expensive to go around a park than through, and if that's all it meant, the statute might as well not exist." Like many Memphians, the Supreme Court felt that there was an intrinsic value in Overton Park that must be protected. If only considering financial expediency, wrote Justice Marshall, the decision to route Interstate 40 through the park was simple. But highlighting the value that the park had to the community—the value that Congress explicitly sought to safeguard—made the decision about more than economic convenience.[3]

Citizens to Preserve Overton Park v. Volpe (1971) remains an important judicial decision, even outside of Memphis. It continues to be considered one of the more environmentally supportive Supreme Court decisions in American history, and it also helped set a precedent for the types of review that courts were permitted to make regarding administrative agencies.[4] Its impact on a national scale was tremendous. "It was at one point—and may still be—the most cited case in administrative law textbooks," notes Newman. *Citizens to Preserve Overton Park v. Volpe* certainly earned its status as a landmark Supreme Court decision, one that will continue to be studied and cited for years to come.

After earning such a monumental victory in the Supreme Court, Citizens to Preserve Overton Park was elated. Their fight, however, was not yet finished. In some ways, it had only just begun. The case was sent back to Judge Bailey Brown in Memphis for a rehearing, and the process of arguing over feasible and prudent routes resumed. This time, CPOP and Newman used a newly created piece of legislation to further boost their argument. The National Environmental Policy Act, enacted on January 1, 1970, required the government to examine closely the environmental impacts of federally funded projects. After examining these environmental effects, the law stated, the government must then disclose the information to the public. Many scholars thus dub this legislation the "Magna Carta" of environmental policy because it requires the government to be transparent in actions that affect the environment and does not afford the government absolute autonomy over all environmental decisions. In any case, it was a major boost to CPOP's cause. According to Newman, the law "played right into our hands because it theoretically forced the state and federal government to do thorough studies of alternative routes and their impacts, so it provided us with more information, more depth." Now with two separate pieces of legislation that required the government to explore alternative routes, Citizens to Preserve Overton Park stood on strong legal footing that would ultimately sustain them through the dispute.

When pressed by CPOP and Judge Brown, highway officials began proposing alternative routes during the trial. These routes were, however, strategically chosen to be unappealing. "You could literally go just north of Overton Park or just south of Overton Park," explains Newman. "They already had plans to build a circumferential highway clear around town. We said going around to the north on the already-planned circumferential was an alternative—it was sometimes referred to as the 'no-build' alternative" because it would require no extra construction. These choices were both feasible and prudent. But the highway department proposed routes that were disruptive and offensive. "The alternative they focused on," Newman says, "was one just south of Overton Park. It went through the St. Peter's Orphanage, various churches, valuable residential areas, and," he laughs, "the federal judge's house." They chose, in other words, alternatives that were easy to eliminate from discussion.

Judge Brown saw through these impractical proposals that were disguised as true alternatives. "After several weeks of trial," Newman remembers, "the judge finally announced that ... it was clear that the secretary of transportation had not applied the rule that the Supreme Court had announced." After making this decision, Judge Brown decided that Secretary Volpe must reconsider the proposed route and its alternatives according to the rules determined by the Supreme Court.

Complying with Judge Brown's order, Secretary Volpe reexamined the alternative routes offered by the Tennessee Department of Transportation and the Federal Highway Administration to determine whether running Interstate 40 through the park was their only real option. Eventually, he gave his decision. Newman explains that Volpe told the highway officials that they "had not proven that there were no feasible and prudent alternatives." Telling highway planners to "go back to the drawing boards," Secretary Volpe demanded that other alternatives be presented before construction continued. Approximately one year later, highway officials went back to Secretary Volpe with a new proposal, one that involved a trench "so that the highway would be depressed as it went through the park, going just south of the Zoo." When they presented this alternative, expressway advocates argued that a trench through the park was not feasible and claimed that they must proceed with the original route through the forest. Volpe disagreed. "He said, 'You still have not established to my satisfaction that there is no feasible and prudent alternative,'" recalls Newman.

This same pattern would continue throughout the decade, testing the resolve of Citizens to Preserve Overton Park. The Tennessee Department of Transportation and the Federal Highway Administration continued to offer radical ideas, proposing deeper trenches and even covered tunnels to no avail. "If they had done a reasonable job of studying alternatives, they might well have been able to build a highway through midtown along one of those alternatives," says Newman. "But they never really made a pass at it. Their professed efforts were so transparently not in good faith that it played right into our hands." Five successive secretaries of transportation—who, according to Newman, deserve a lot of credit for the way they handled the situation—repeatedly refuted the state's claims that there were no feasible and prudent alternatives to building through the park. Eventually, state highway officials unhappily settled with routing the interstate north of Memphis along the circumferential expressway.

Against all odds, Citizens to Preserve Overton Park stopped the expressway from destroying their public space. To this day, their decades-long battle and eventual victory show the power of grassroots activism in shaping government decisions. CPOP's success is even more remarkable when considering the breadth of opposition they faced throughout the controversy. In the courtroom, these citizens squared off with federal and state officials who fought diligently to run an expressway through Overton Park, but the ferocity of the fight in the courts paled in comparison to the aggression used outside of the legal arena.

As the controversy was prolonged, many businesses, newspapers, and other news agencies became outspoken advocates of routing the interstate through Overton Park, even going so far as to publicly ridicule Citizens to Preserve Overton Park. According to Newman, local newspapers "put out editorial cartoons heaping scorn on the Citizens to Preserve Overton Park, who were portrayed as being unrealistic, little old women in tennis shoes and so forth." But the rhetoric used by local newspapers seemed civil compared to the words and actions of the trucking and transportation industry. "The trucking industry had, and still has, a newsletter called *Transport Topics*. In this paper, they were just brutal in their attacks on our clients [CPOP]." In an article entitled "Environmentalists Spilling Blood," the newsletter's associate editor wrote:

> Just as the rabble-rousing street mob clamored for the head of Jesus in Jerusalem nearly 2,000 years ago, so have the environmentalists who—through lengthy litigation—blocked construction of Interstate 40 here spilled the blood of innocent victims. . . . The spilled innocent blood is that of those unfortunate individuals killed on overcrowded Memphis

streets—who would still be alive if the expressway had been built through Overton Park, as it was originally scheduled to be constructed years ago. As a memorial to this needless death and suffering, perhaps the self-appointed environmental messiahs might wish to perpetuate the memory of those innocent people they have martyred while saving the Overton Park trees. If so, it might be done in this way: They could tie red-blood ribbons around the trees corresponding to the number of people who have been unnecessarily killed while construction of the highway has been blocked.[5]

With these estimated, unverified, strategically concocted assumptions of CPOP-caused deaths, the transportation industry mercilessly attacked Citizens to Preserve Overton Park. Personal attacks such as these make CPOP's resolve, commitment, and passion—and eventual victory—even more commendable.[6]

Taking their case to the highest court in the nation, Citizens to Preserve Overton Park used the full extent of the law to save a space full of nature, culture, and opportunity. Their persistence for another decade is perhaps even more remarkable. Even with ruthless attacks from local media and national transportation industries, this small group of Memphians persevered. Citizens to Preserve Overton Park's success in the expressway controversy stands as a testament to the effectiveness of grassroots activism and reveals the power and beauty of democracy. With unwavering diligence and intelligent resistance, these citizen-stewards preserved Overton Park.

A POST-VICTORY BOOM

In the wake of CPOP's historic victory, Overton Park's popularity soared. People flocked to the park for all kinds of reasons, eager to enjoy Memphis's most vibrant public space. Children continued to take advantage of the park's playground equipment as they had for years, while teenagers and adults attended the Academy of Art, listened to sold-out shows at the Shell, and jogged through the Old Forest. Others planned community events in the park, excited to gather with their neighbors over exercise, education, and family fun. Following the Supreme Court case, Memphians of all ages, races, and genders embraced Overton Park and claimed it as their own. According to Willy Bearden, a Memphis author and filmmaker who has worked extensively on documenting Overton Park's history, "People had an affinity for this place."[7] The 1970s were, in some ways, a peak in the park's history.

As it had since the 1930s, the Overton Park Shell played a pivotal part in making Overton Park a popular destination during the 1970s. After moving to Memphis in 1971, Willy Bearden plainly remembers attending lots of concerts at the Shell: "In 1971, it was right in the middle of the hippie times, and believe me, I was right in the middle of all that." Rock bands, soul singers, R&B stars, and funk musicians headlined shows at the Overton Park Shell. "I saw the Allman Brothers there. I saw Trapeze. And Trapeze even has a famous album cover that they took from the [Shell] stage of the crowd. . . . I saw Steve Cropper of Booker T. and the M.G.'s play there, and Eddie Floyd, and just all these incredible musicians." Other notable performers during this era included the Marshall Tucker Band, Black Sabbath, the Bar Kays, Seals and Crofts, Isaac Hayes, Rufus Thomas, Carla Thomas, and Deep Purple. For many Memphians, these high-profile concerts at the Shell made Overton Park an attractive place. Bearden speaks for both himself and others when he describes Overton Park as "that place where you could go and meet people, where you could go and hear great music."[8]

On occasion, these concerts even touched Bearden on a personal level. He remembers attending a show in the early 1970s with a close friend who passed away shortly before Bearden was interviewed for this book. "When you get a little older, you start losing friends. That's tough," he says. On the day of his friend's funeral, Bearden saw a picture of himself and his friend at this Shell concert, and immediately, the memory rushed back into his mind: "I remember that moment like it was fifteen minutes ago. I remember what we were saying, how we were dressed, what we were doing. It's just a wonderful thing, and it was because of the park."[9]

But as it has throughout its history, Overton Park at this time meant something different for each individual visitor. Whereas Bearden and his friends loved rocking out at the

A large crowd enjoying a concert at the Shell. Photo courtesy of William Bearden, author of *Images of America: Overton Park.*

Shell during the 1970s, others valued the park for its more peaceful aspects. Sally Jones Heinz grew up just west of the park and remembers taking summer classes at the Memphis Academy of Art as a teenager: "It was 'Saturday School'—that's what it was called. I had a good friend who lived on Buena

Vista right next to the park, so I would walk from my house to her house, and then we'd walk to the [Academy of Art]. She took her class and I took mine, then we'd walk home." Making this trek became habitual for Heinz, and after several summers of following the same path, she became even more intimately attached to the park. "I knew that sidewalk and every crack in it," she laughs. "It was a really nice walk to take."[10]

Given that she took Saturday classes at the Academy of Art for several summers, Heinz's interest in art was apparent. She focused mostly on pottery, though she admits that she never became a great sculptor. Heinz had another reason to be so

The main building of the Memphis Academy of Art (later renamed the Memphis College of Art). Photo courtesy of William Bearden, author of *Images of America: Overton Park*.

closely involved with the Academy of Art, though. Her father, Jameson Jones, was president of the Memphis Academy of Art for ten years after serving as the dean of Southwestern (now Rhodes College) from 1955 to 1972. And just like his daughter, who walked to all of her Saturday classes, Jameson Jones walked to work at the Academy of Art every day. "Walking was just a big part of his life," she explains, and it continued to be important for him even after he stopped working. "When he retired, he continued to walk. I think his life would have been missing something if that park had not been there for him to walk in," Heinz says. In his final years, he moved into the Parkview, which is a grand hotel–turned–retirement home on the edge of Overton Park, because he wanted to continue his walks. "You know," Heinz reflects, "I don't think he would've gone anywhere else because he had to be at the park."[11]

Jameson Jones was just one of many Memphians who found pleasure in using the park for fitness and relaxation. Aside from the amenities that the park offered, countless people took advantage of the park's natural elements as a way to exercise and relieve stress. Blanchard Tual, a native Memphian, made a point to run in Overton Park because doing so enabled him to engage his body while disengaging his mind. "I literally ran in Overton Park five or six days a week from 1975 to 2005," he says. "I mean, five miles a day, five or six days a week. I loved the park, and I would look forward to that one time I could just be off by myself and have some peace."[12]

Gigi Wischmeyer grew up near the park and was often there during the 1970s. Like Tual, she too used the space for exercise and relaxation. "When I was about thirteen, so maybe 1972–1973, there was a guy named Jim Migdoll who would teach free yoga for all comers," she remembers. "He had a little collection basket that we would put fruit in, and that was his only payment." The yoga classes took place on the edge of the Greensward, the largest open green space in midtown Memphis. While Wischmeyer and her family practiced yoga just to the east of Veterans' Plaza and the Doughboy Statue, others threw Frisbees in the heart of the Greensward and enjoyed the wide expanse of grass. Echoing Willy Bearden, Wischmeyer explains that such events were crucial to the "hippie" culture that the park sometimes facilitated in the 1970s.[13]

Fruit-financed yoga classes were not the only community events held in Overton Park during this decade. Jimmy Ogle, who often visited Overton Park when he was a student at neighboring Southwestern, served in an administrative capacity for Memphis parks in the late 1970s, and he organized several events in Overton Park in this role. From citywide bicycle derbies to 5k runs, Ogle tried to bring family-oriented events into the park. Chief among these events was the Memphis Fun Fest, which began in 1979. "Fun Fest was in the Greensward area and Rainbow Lake area of Overton Park," he says. "We had arts and crafts vendors out there and lots of different games." Aiming to provide recreation opportunities for Memphians of all ages, Ogle and others worked hard to create a fun atmosphere in the park.[14]

The boom that ensued following the 1971 Supreme Court decision in *Citizens to Preserve Overton Park v. Volpe* can be seen in many different ways. The Overton Park Shell soared in popularity among music enthusiasts, while the Academy of Art continued to nurture and refine artistic interests in teenagers and students alike. Overton Park also increased in attractiveness to those who sought to exercise and free their minds. The park in the 1970s solidified itself as a major destination for Memphians. From the perspective of an early 1970s park user, continued success and popularity seemed to be part of Overton Park's future.

THE START OF A SLOW DECLINE

Despite all of the signs that the park would perpetually thrive, Overton Park began to slip into a slow decline around the middle of the 1970s. In the aftermath of the 1971 Supreme Court decision, the park certainly increased in popularity, but as Charlie Newman points out, the battle did not end with the court's decision, and the fact that more Memphians than ever were now enjoying the park was not enough to deter highway officials. The Tennessee Department of Transportation and the Federal Highway Administration were still confident that Intestate 40 would be built through Overton

Park. This high level of confidence among highway planners certainly impacted city officials' duty to take care of the park: why should they pour resources into a public space if they were being led to believe that it would soon convert into an expressway? Thus, the boom that occurred after CPOP's victory was short lived. Soon after the park's rise in the early and mid-1970s, Overton Park began to deteriorate. A general perception of Overton Park as unsafe and unkempt began to discourage visitors.

Charlie Newman, who had a close relationship with Overton Park during this decade, feels strongly that the park was abandoned by most city officials in the 1970s. "I think Overton Park was neglected during those years somewhat intentionally because the city government had no desire to enhance it." If they made the park a more attractive place, he reasons, highway officials would have encountered even more resistance from park-loving Memphians. Of course, this characterization of apathy cannot be applied blindly to all city leaders in this era. As mentioned above, Jimmy Ogle and the Recreation Department worked hard to provide opportunities for people in Overton Park. For the most part, however, Newman's belief that the park was neglected by city leaders rings true.[15]

One of the most powerful illustrations of this neglect actually occurred outside the park's boundaries. The destruction of hundreds of private homes leading up to the park reveals the federal, state, and city government's attitudes in this decade.

Aerial view of the widespread demolition just outside the park as
Interstate 40 was being constructed. Photo courtesy of William
Bearden, author of *Images of America: Overton Park*.

Fully expecting the interstate to be routed through Overton
Park, highway officials destroyed 408 homes, 84 duplexes, 266
apartments, 44 businesses, 5 churches, and a fire station that
lay in the road's path.[16] Given that they had already made
the financial investments to destroy these homes, businesses,

and churches, government leaders had no intention to invest in improving the park. Neglecting the park would only help justify their desire to use it as a route for the interstate, while tearing down the buildings leading up to it would provide yet another reason to pave it.

Several people remember the destruction of these buildings along the proposed expressway path and recall the eeriness of being surrounded by vacant lots. Martha Kelly, a Memphis artist and daily park user, currently lives in her grandmother's old house, which is within walking distance of the park. "The whole time I was growing up—the 1970s, 1980s—there were empty lots on both sides of this house," she says. The home she now lives in was barely spared. Ernest Kelly, Martha's father, expands upon his daughter's memory, explaining that an entire block had been torn down leading up to Martha's current home.

Martha's and Ernest's memories are joined by a host of others. Sally Jones Heinz, who lived two blocks from the zoo, says that when she was a child, "houses were being torn down only one street away. I had friends displaced, friends who had to move." Similarly, Gigi Wischmeyer remembers passing by demolished properties frequently: "Every day, I would ride my bike from Morningside Place down Williford Street to school. During that time, all those properties had already been torn down for the expressway to go through, and they sat dormant for years and years." This widespread destruction of buildings stands as a poignant reminder of the indifference that many leaders felt toward the park during this decade.[17]

The destruction of nearby homes and businesses, along with more subtle deteriorations within the park itself, caused Overton Park's quality to decline, and this neglect had a lasting effect. Heinz elaborates on this negative perception: "When I was in college [in the late 1970s], that was probably the bleakest period for the park in terms of its image." The excitement and entertainment that had characterized park activities in the previous years slipped away. Instead, the park was now "full of people waxing and cleaning their cars on the weekend. I think there was even a streaker." Drug deals and prostitution occurred in the park, too. Such rumors made some people feel uneasy, but not Heinz. She and her friends continued to run through the Old Forest, and her dad maintained his tradition of walking in the park. But for years, the space felt drab and desolate to Heinz and other Memphians. Overton Park's signature vibrancy was gone.[18]

CONCLUSION

The 1970s in Overton Park began with a surge of success. With their dramatic upset in the Supreme Court, Citizens to Preserve Overton Park saved one of Memphis's most important public spaces, and these citizen-stewards maintained their efforts for a decade to ensure that Overton Park would never fall victim to aggressive highway bureaucracies. In the wake of CPOP's 1971 victory, Overton Park entered into what was a pinnacle in its history. Memphians streamed into the park for all sorts of reasons. They jammed with the Marshall Tucker Band and Isaac Hayes, sculpted pottery at the Academy of Art, and paid for yoga lessons with apples and bananas. The park's popularity was evident to all who visited. A dynamic location in the geographic heart of Memphis, Overton Park was a cultural, social, and natural haven within what was then a struggling city.

But Overton Park's luster soon dimmed. As the 1970s wound to a close, so did Overton Park's perception as a place where all Memphians wanted to be. It became, as one longtime park user says, a "sadder place."[19] The confidence of highway officials and city leaders helped drive this downward trend. Fully expecting an expressway to be built through the park and discounting CPOP's ability to continue waging a successful fight, transportation officials destroyed hundreds of buildings that stood in the expressway's proposed path, making Overton Park seem like nothing more than a speed bump in a near-finished project. In the process, these officials also fueled blight and impoverishment in neighboring Binghampton, a community just east of the park that still faces socioeconomic struggles. Between destruction outside of the park's boundaries by state and federal entities and neglect within by city leaders, Overton Park's quality suffered. The sense of excitement and energy that park users had nurtured and strengthened in the early 1970s faded. Overton Park stopped being thought of as a multifaceted urban sanctuary.

This negative perception plagued the park for years, lasting through the end of the twentieth century. But despite the overall view of the park as a less vibrant institution, many Memphians continued to visit. They challenged Overton Park's lackluster stereotype and kept using the park. Although it was less glamorous during the 1980s and 1990s, Overton Park was embraced by a faithful contingent of Memphians who still saw its value and potential.

CHAPTER 4

THE TWENTIETH CENTURY CLOSES
1980–2000

O verton Park reached what was perhaps its lowest point during the last two decades of the twentieth century. The perception of Overton Park as a vibrant gathering place dimmed, and although the park was not abandoned, it had lost some of its appeal. It became a place riddled with violent crime and other illegal activity, a place that many Memphians perceived as dangerous. In the late 1990s, new professors at nearby Rhodes College were even warned not to enter the park because, according to campus safety leaders, it just wasn't safe. "I don't think it ever died, if you will," explains Janet Hooks, who recently served as director of parks and neighborhoods for the City of Memphis, but Overton Park certainly garnered less positive attention and became less welcoming.[1]

Though Citizens to Preserve Overton Park had been triumphant, the interstate controversy can be viewed as one of the factors that led to the park's decline. Willy Bearden, who has used the park extensively since the 1970s, bought a house in 1984 in a nearby neighborhood that had been deeply affected by the interstate. Many of the homes were demolished in anticipation of the expressway running through Overton Park. Noting the widespread destruction and the effect it had on the park, Bearden reflects on a stroll with his daughter soon after they moved into their new home. They walked past vacant lots and across empty driveways and abandoned walkways. Out of nowhere, Bearden's three-year-old daughter paused and asked: "Why are there sidewalks to nowhere?" Bearden was stunned. Even a three-year-old recognized the destruction that plagued some of the areas surrounding Overton Park. These "sidewalks to nowhere," though technically outside the park, were indicative of the neglect within.[2]

Despite its somewhat diminished status during the 1980s and 1990s, Overton Park did not become obsolete. Because of the newly formed groups Park Friends and Save Our Shell and institutions like the Memphis Academy of Art, the Memphis Zoo, and the golf course, the park endured through one of its most difficult periods. Events organized by the Recreation Department helped bring Memphians into the park, too. And, more so than at any other time in its history, the park was

"Sidewalks to nowhere." Photo by William Bearden.

embraced by runners. With cooler temperatures, softer surfaces, and abundant shade in the Old Forest, the park offered
the perfect running location in the heart of Memphis. Though
it may not have been the jewel it was in the 1970s, the park was
still an important place for Memphians—and after the turn
of the century, it would soon become a jewel again.

RUNNERS EMBRACE OVERTON PARK

Running had been a popular activity in the park before the 1980s and 1990s. As noted earlier, Blanchard Tual and others ran through Overton Park several times a week during the 1970s, using the park for both its trails and its ability to facilitate relaxation. Because he loved this space so much, Tual maintained his tradition of running in Overton Park for thirty years. But during the 1980s and 1990s, the park began to garner even more attention among running enthusiasts in Memphis for several reasons. Whether running on the golf course or through the forest, people loved the soft surfaces the park provided because they made for less wear and tear on knees and ankles. Temperature was important as well. People familiar with the forest constantly comment that, during the summer, the Old Forest is at least ten degrees cooler than the surrounding streets. Because of such features, a running community developed within the park. Whether crossing paths on the perimeter of the golf course or running the trails through the Old Forest together, Memphians became acquainted with both the park and one another through their runs.

Melanie White, who has become a cornerstone of the Memphis running community, helped establish Overton Park as a running destination in the 1980s. Moving to Memphis in 1985 from San Francisco, White immediately began running in Overton Park. "Running has been how I was introduced to the park," she explains, "and how I became a park advocate." Before trails were installed in the Old Forest, White ran on the roads through the park. She met up with friends at the golf clubhouse, and they would loop through the forest and by the Brooks Museum. While running this loop in the mid-1980s and through the trails at later points, White cultivated some of her most important relationships: "I've made my very, very best friends in the park. I met people while running who have become my best friends."

Some of White's most important friendships were forged unintentionally. She met one of her closest friends, Lyn Reed, while running through the park alone. "Lyn and I ran past each other about three times one day," she recalls. "And finally, we just stopped." Striking up a quick conversation, the two agreed that, if they ever wanted a running partner, they would call one another. "So I called her," says White, "and we started running. We ran together every morning for years."

Other friendships were more casual. White ran with a variety of people in the park, from professors at nearby Rhodes College to employees at the *Commercial Appeal.* These runners-turned-friends came from all over the city and from different racial, social, and political backgrounds. White, who still runs in the park to this day, says that these relationships made Overton Park feel like her second home.[3]

Like White, Gigi Wischmeyer also felt a strong sense of community when running through the park. Her community even expanded beyond fellow runners. "I've run in the park since 1990—four days a week, every week. The people that I would see, regardless of their mental health, their socioeconomic status, or where they slept that night, became a community," she says. "I would recognize the same people every time I went. They would always wave and say hi." One friendship in particular stands out in her mind. "There was one man, and I'm so embarrassed that I don't remember his name. . . . He'd always chat with me, and I always ran with my dog. One day when I saw him, I didn't have my dog. He said, 'Where's your dog?' I told him that he had been hit by a car. The man stopped and started crying, saying, 'Please don't tell me that. Please don't tell me that.'" The frequent interactions between Wischmeyer and the man fostered a sense of community for each of them, and they began to rely on one another. While some people questioned her for running alone through the park among strangers, Wischmeyer never felt in danger. "I actually felt like those people were looking out for me," she says, just as friends should do.[4]

Michael Cody, who is perhaps the most well-known runner in Memphis, started running more in Overton Park in the early 1990s, too. While he has made many friends when jogging through the park, his primary motivation for running there in the 1990s was the soft surfaces provided by the golf course and Old Forest. "Up until about 1993, I guess, I had run probably thirty-five to forty marathons and thirteen to fourteen Boston marathons," Cody explains. "I was sort of a fanatic. I was running hundred-mile weeks, all of it on the pavement." Eventually, all the miles on asphalt began to take a toll on his body. "I more or less wore my heel pads off. I didn't have any cushion in my heels. They would split and bleed," he says. At that time, Cody was living in a small apartment across Poplar Avenue from Overton Park. Knowing he needed to find an alternative surface if he wanted to continue running, he began to explore the park. "I just started going out my door, crossing Poplar, and basically just ran on the golf course," he laughs. Soon, Cody realized that his runs sometimes interfered with the golfers. When he recognized

Michael Cody, left, running through Overton Park.
Photo courtesy of Michael Cody.

this, he began exploring the trails in the Old Forest. The trails were not well-maintained at that time, and poison ivy and downed branches were rampant. The trail system was archaic compared to the limestone loop the park boasts today. But between unkempt trails and lush fairways, Cody found softer surfaces that enabled him to run without pain.[5]

Arguably, running was among the most important park activities in the 1980s and 1990s, aside from happenings tied to park amenities. Jogging through the forest, along the park roads, and on the golf course, runners solidified Overton Park as a fitness destination. And as they wound along trails and roads, they nurtured a sense of community with each other and with those they ran past. Runners helped both their fellow fitness enthusiasts and complete strangers feel welcome within this public space. Because of this, they increased the health of Overton Park while also nurturing their own physical wellbeing.

PARK INSTITUTIONS KEEP THE SPACE ALIVE

Like the Memphians who ran in Overton Park during these decades, park institutions helped invigorate the park during a lull in its popularity. The Memphis Zoo and the golf course provided Memphians opportunities for recreation and stimulation, bringing animal lovers and golf fanatics into Overton Park. The Memphis Academy of Art, which was renamed the Memphis College of Art in 1985, also inspired people to come to the park. Instilling a passion for art in both full-time college students and high school students who took classes during the summer, the College of Art made a lasting impact on some Memphians. And though there are fewer memories of this amenity, the Brooks Museum was important as well. Park institutions helped keep Overton Park alive in the final decades of the twentieth century, providing a much-needed spark to the park during the 1980s and 1990s.

Founded in the early twentieth century, the Memphis Zoo had long been one of the most popular amenities within Overton Park. But the zoo was in rough shape by the 1980s. Its cages were cramped, its infrastructure aging. During the interstate controversy of the 1970s, the zoo's overseers had made few improvements out of fear that it might be closed or relocated as a result of the interstate, explains Jimmy Ogle, former deputy director of the Memphis Park Commission and current Shelby County historian. The zoo's condition suffered immensely throughout this time. By 1980, Ogle says, "it was miserable."[6]

After the interstate was finally defeated, improving the zoo took on new interest. James Jalenak, who used to handle much of the zoo's legal work, was part of this revitalization effort in the early 1980s. "A group formed that got to be known as the Breakfast Club—we got to be known by that nick-name because we met every Wednesday morning at Shoney's

Restaurant." In addition to Jalenak, who was a member of the Memphis Zoological Society at that time, the Breakfast Club was full of well-intentioned Memphians from the business and civic community. After a few meetings, the group established a lofty goal: they wanted to build an entirely new cat facility inside the Memphis Zoo.

"Our slogan," says Jalenak, "was 'Free the Cats.'" And because the cats in the zoo lived in such deplorable conditions, this phrase was not taken lightly. "[At that time,] what is now the Cathouse Café was in fact the Cat House," he explains. "It was an old-fashioned, steel bars, concrete floors, terrible-smelling cage of big cats." During this era, people were realizing that zoos needed to be more animal-friendly and more focused on conservation and education. These facilities were no longer thought of as mere entertainment destinations. So with their new goal in mind, the Breakfast Club began working to provide a better habitat for the cats that lived in the Memphis Zoo.

The Breakfast Club's work primarily consisted of fundraising, and they began by reaching out to the City of Memphis. "We proposed that if we raised two-thirds of the money," Jalenak says, "the City would provide a third of the money." City officials agreed. Jalenak and others then worked on returning the land near the zoo that had been given to the State of Tennessee for interstate construction back to the City of Memphis. Once they got this property back, they would more easily be able to complete their project. And after successfully regaining the property rights to the land, the zoo made its next move, hiring a fundraising consultant. At this point, their plan changed.

After speaking with their consultant, Leo Arnoult, Jalenak and the others learned that their plan was not ambitious enough: "Leo conducted a study to see how the prospects were for fundraising. He came back and said, 'You're not going to be successful because the project is too small. You've got to make it bigger—people will give to something big, but they won't give to something small.'" With Arnoult's advice, the Breakfast Club expanded their goal. Ultimately, they ended up creating an entirely new entrance and front plaza, converted the Cat House into the Cathouse Café, and created Cat Country, a new exhibit for the zoo's big cats. All of this work was completed by 1994. Throughout the rest of the decade—and still today—the zoo has continued to improve its facilities. Other facilities added during this time showcased farm animals, komodo dragons, and primates, all of which boosted the zoo's quality and popularity. With the construction of Cat Country and other new amenities, the Memphis Zoo began the ascent to becoming a world-class institution.[7]

Of course, Jalenak, the Breakfast Club, and the Memphis Zoological Society played a tremendous role in improving the zoo and thus improving Overton Park. Fundraising became one of the most important ways of improving entities like the Memphis Zoo, and, as Jalenak notes, the fundraisers were delighted to see the result of their efforts and to know that the lions, tigers, primates, and farm animals could now lead better lives. But the impact of these new facilities was perhaps most appreciated by the people who cared for the animals on a daily basis. Retired zookeepers Kathy Fay and Richard Meek remember clearly the transition of the Memphis Zoo from old-fashioned to state-of-the-art.

Richard Meek began working at the Memphis Zoo in 1974. A former high school biology teacher in rural Arkansas, Meek became the zoo's new curator of education and served in this role for five years. He left in 1980 to pursue other work opportunities, but by 1987, he could no longer resist the urge to come back. Born and raised on a small farm, Meek was able to continue his passion of caring for animals when he returned to the zoo in the late 1980s to work not as curator of education, but as a zookeeper.

Kathy Fay took a completely different path to the zoo. She grew up in the Chicago suburbs, she explains, and when she was twenty-nine years old, she decided that she wanted to work in a zoo. At this point in her life, she was working in a pet store in Minneapolis. Having loved animals all her life, Fay decided that she wanted a more hands-on career with a variety of animals. Determined to work in a zoo, she began the job search. "I went to the library, and I looked in a book called *Zoos and Museums in the U.S.* I wrote down one hundred addresses, and I sent out one hundred résumés. I got a telephone interview with the people here at the Memphis Zoo, and they hired me over the phone." Knowing almost nothing about the city, Fay moved to Memphis in 1987 to start her new job as a zookeeper.

As fate would have it, Fay and Meek met in 1987, when they both took their new jobs at the zoo. "We were swing keepers," Fay explains, which meant that they might work in different exhibits with different animals each day. "It depended on who was on vacation, who needed extra help, and if someone called in sick. And it was so amazing to have that variety." But on days when no one was on vacation and no one called in sick, swing keepers were put on special projects. "So Richard and I got thrown into the feed storage barn to clean it out, which probably hadn't been done in fifteen years." The work, Meek and Fay remember, was horrible. "We had to move hundreds of bales of hay, move the pallets that were underneath them, and clean up all the unpleasantness that was underneath those pallets and had been there for years—and that includes rats and mice," explains Fay. But the company made the hard work well worth it. "I had never met anybody like her," Meek says. "I really hadn't. And here we were, doing this unbelievably un- pleasant job and dodging brown recluses while we were at it. I watched somebody—this little tiny woman—work beside me all day long, and we talked about everything you could possibly imagine, from literature to science to politics." Immediately,

Meek was enamored. They were both in other relationships at that time, but years later, they each found themselves single. "And the rest . . ." Fay begins, ". . . is history!" finishes Meek.

In addition to seeing their relationship blossom while working at the zoo, Fay and Meek watched the space around them flourish as well. "It was not quite an embarrassment when I started at the zoo [in 1987]," says Fay, "but it was an old-fashioned zoo," with, Meek adds, "a lot of concrete and iron bars and so forth." But within a few years of their meeting, the zoo began to change. Like Jalenak, the two keepers remark that Cat Country vastly improved the lives of the Memphis Zoo's big cats and identify this moment as the beginning of the zoo's transformation. Fay and Meek, however, were more touched by another improvement. When Primate Canyon opened in the mid-1990s, they were moved by the actions of two siamang monkeys entering the new exhibit for the first time.

The two keepers remember clearly the day these two siamangs relocated into their new home. Meek recounts the story: "Debbie and Danny had lived in a long, concrete and metal cage where they swung back and forth all day on bars and did that wonderful siamang yell that you can hear from miles away," he says. When Primate Canyon was finished, keepers from across the zoo worked to get the animals ready for their move. Meek recalls Danny and Debbie's first moments in Primate Canyon, the first time they could get out of their concrete and metal cages. "We let them out, and they touched grass for the first time in their entire lives," says Meek, his eyes watering. "They came across the rope, looked down, and dropped to the grass. They sat there in the grass just feeling it and running their hands over it. We all sat and cried like babies."

In that wonderful moment, Meek knew the Memphis Zoo had turned the corner from an antiquated institution to a

Kathy Fay with Ty the elephant. Photo courtesy of Kathy Fay.

modern-day zoo. "That was the day zookeeping went from being something really special to being something magical," he says. "The day Debbie and Danny touched grass was a highpoint of my life."

Fay, like Meek, shares clear memories of the zoo's transition to a more animal-friendly institution. She specifically recalls a new training method that was adopted for Ty—short for Tyranza—an African Bush elephant. "She had been a circus elephant, and then she came to the zoo. She had always been worked with what they call 'free contact'—when a person goes in with a stick with a hook on the end and dominates the elephant, telling her what to do. And the elephant does it because if she doesn't, she's going to get in trouble." Fay remembers that she was taught to train Ty that way, and she used this method for years.

Zoos across the country began transitioning to a new training method during the 1990s, one that was safer for the keepers and less oppressive for the elephants. "We gradually transitioned to something called 'protected contact,' which is where the elephant is behind a barrier and the keeper and the elephant are separated by this barrier." While making this transition, Ty realized that, without a stick to fear, she no longer had to listen to the keepers—she now had the option to turn, walk away, and ignore her trainers. Fay explains that this new method was "wonderful psychologically for Ty and wonderful psychologically for us, too, because we hated dominating them. But it was a little frustrating when you were trying to give the elephant a bath or something."

After some time passed, Ty began to trust Fay, building an elephant-human partnership. "There was a behavior called 'sit,' and it was like a dog sitting. She decided she wasn't going to do that anymore and had gone a month without doing the behavior." Finally, after weeks of failed attempts, Fay tried a new approach: "One day, I asked her to sit, and I had this bucket of fruit hanging from my belt. I pulled it off and said, 'Ty, sit.' And she looked at me, and then she looked at the fruit. I was telling her, 'If you sit, I will give you this bucket of fruit,'" explains Fay. "What blew me away—and still blows me away—is that she understood what I was saying to her. I was making a promise. She trusted me. She sat, and that was the first time she had sat in a month. I dumped the bucket of fruit in front of her . . . and she sat every day after that." Through mutual trust and respect, Ty and Fay forged a cooperation that lasted for years.[8]

With these new improvements in both the treatment of animals and their living quarters, the Memphis Zoo offered a more holistic, beneficial experience to its inhabitants, workers, and visitors. And by doing so, it helped sustain Overton Park itself. By bringing so many people into its now-blossoming institution, the Zoo ensured that Overton Park remained relevant in a period of diminished excitement.

Like the Zoo, the Memphis College of Art helped bring enthusiasm into Overton Park during the 1980s and 1990s.

It was especially helpful because it brought young, creative people into the park. At the College of Art, full-time college students and high school students taking summer courses explored their artistic passions and worked to hone their skills. Martha Kelly, a successful Memphis artist, was one of the students who took advantage of what the College of Art had to offer. Through attending classes and painting in Overton Park as a high school student, she realized that she ultimately wanted to pursue a career in art.

Her passion for art began when her father, Ernest Kelly, enrolled her in a few summer classes as a teenager. "I think I was fourteen," she says, "and I was painting that little birdbath over at the Brooks Museum that has the little child kneeling in it." Using water colors, Kelly became immersed in what she was doing. After working on her painting for a while, she was overcome with a sense of joy and conviction. In that moment, she remembers, "the paints kind of swirled on the page in front of me and did something that I didn't expect. It was magical, and it was the first time where what I had painted was better than what I saw in my head." After this experience, Kelly knew that she wanted to be an artist. "I was hooked. I was like, 'Wow! That's it. That's what I want to do.' And ever since then, I've been working toward being an artist."

Kelly's experiences at the College of Art were not limited to the revelation of her artistic passion. She continued to take classes to improve her abilities, determined to become a successful painter. "I went to a high school with a bad art program, so I took a couple summer classes at the College of Art just to audit," she explains. Auditing these classes helped her tremendously. "I wanted to be ready. I knew I wanted to be a studio art major." Although Kelly attended a small college in Kentucky for her formal education, she credits the College of Art for preparing her to pursue an art major. The summer courses she took as a teenager enabled her to succeed once she enrolled in college.[9]

For Kelly and other aspiring artists, the College of Art was important for developing artistic talent, and because it brought so many young people into the park, the College of Art was also essential to improving the park during the 1980s and 1990s. It encouraged Memphians with creative dispositions to spend time in Overton Park and helped the park strengthen its status as a cultural hub in the city.

On a more informal level, the golf course in Overton Park helped attract Memphians into this space as well. A nine-hole course that runs throughout the park, the Links at Overton Park provided an afternoon or morning of fun for both casual and serious golfers. Although not accounted for in George Kessler's original plan for the park, the golf course was easily implemented a few years after because of the vast amount of open green space. Cherished by Memphians in 1906 and the 1980s and 1990s, the Links at Overton Park has remained a favorite park institution for many patrons.

Willy Bearden was one of many who enjoyed rounds of golf in Overton Park. In the mid-1980s, Bearden began playing golf with his friends, never having had much experience with the game prior to this point. He is the first to admit that his golf game was not strong. "I was a terrible, terrible golfer," he laughs. But his biggest shortcoming had nothing to do with his swing. "My downfall was that somebody gave me the *Audubon Guide to Trees.* I was just ruined as a golfer then because I'd be thinking, 'What is this, a Persimmon tree?' And they would yell, 'Come on, it's your turn. Putt!'" His friends would become frustrated. Finally, Bearden gave up golfing and focused on other park activities, but he still fondly recalls his golf stories from this era in Overton Park.[10]

Donnie Bailey, who grew up in Memphis and became the golf pro at the nine-hole course from 1986 to 1992, remembers lots of people like Bearden who played in the park solely for enjoyment. "I became friends with many casual golfers. They weren't always the best players in town, but they had fun. And they usually came at the same time every week." Some golfers, however, were more serious. Whether as children or adults, many took to the tees with focus and determination. Bailey explains that several top golfers, including Frank Moore, Buck White, Hillman Robbins Jr., and Bailey's teacher and mentor, W. D. Fondren, played their first rounds of golf in Overton Park. Several of them were caddies at the course, too, carrying bags and clubs for other golfers. The most successful of these young players would go on to win major tournaments and collegiate national championships, and some played in the world-renowned Masters Tournament in Augusta, Georgia.

Other players achieved less success, but they still played golf in Overton Park with passion. And many of them played because of Bailey's hard work. "We revamped the kids' tournament, which was started by my mentor, W. D. Fondren, and his brother, Jake," Bailey says. "I say 'we' because it was much more than just me. I was in charge, but many other people were key to the tournament's and the course's success." Before he began in 1986, numbers had dwindled at the famous competition. But participation soon increased. "By 1992, hundreds of kids were playing in the tournament," he boasts, "and the tournament became a big deal again." Other Memphians made sure the tournament remained important. Famous Memphis golfer Dr. Cary Middlecoff, who won forty PGA Tour tournaments, came to the tournament in the late 1980s to present trophies to the winners, and local television stations filmed the ceremony. Bobby Hall, a writer for the *Commercial Appeal,* gave lots of media coverage to the children who participated, recalls Bailey. For the participants, it was a thrill to see their names and pictures in the paper.

Bailey remembers his time as the golf pro at Overton Park fondly. There were battles, he says, like the time he and others stopped the Brooks Museum and the Memphis College of Art from paving part of the ninth green for a parking lot.

But his years in the park were mostly positive. "We tried to help everyone who walked into the clubhouse," he remembers, explaining that he sometimes worked eighty-hour weeks to do so. Serving the people who came to play on the course was especially important for Bailey. "As a private contractor," he says, "I was a businessman. But I had a public responsibility." He felt this responsibility to serve the people of Memphis partly because of the ultimate sacrifice his twin brother, Ronnie, paid for these same citizens. "Ronnie flew a Cobra gunship in the Vietnam War. In his career, he logged over a thousand combat hours. Thankfully, he came home safely in 1969. When I got the call that he was back in the United States, it was one of the best days of my life." Years later, in 1983, Ronnie was involved in a fatal accident while training in reserve duty. After explaining this, Bailey grows quiet and then says, "I felt a duty to help other people after my brother's death." Enabling others to play the game that they loved, Bailey fulfilled his responsibility, honoring his brother.[11]

The Memphis Zoo, the Memphis College of Art, and the golf course in Overton Park were all essential to maintaining this space through a difficult period in its history. The zoo, through improving its facilities and offering a more modern experience, helped bring countless Memphians and non-Memphians into the park, as children and adults came to see the lions, primates, and elephants the facility proudly housed. The College of Art helped inspire artistic young people to work on their craft and instilled a lifelong passion for art in some students. And the golf course offered an outlet for recreation and competition, even if some golfers were more concerned with identifying trees than improving their scores. These new amenities enabled Overton Park to hold onto some of its vibrancy and resolve through two difficult decades.

THE POWER OF CITIZEN-STEWARDS

In addition to park institutions, Overton Park–focused citizen groups were of utmost importance in sustaining the park near the end of the twentieth century. Groups like Park Friends and Save Our Shell helped protect the park from those who wanted to disrupt it, including some park institutions: Park Friends and the Memphis Zoo started what has become a brutal and lengthy fight over parking on the Greensward, the largest open expanse of green space in the park. Just as important as their tangible accomplishments was the interest in the park that these groups incited. By devoting themselves to Overton Park's wellbeing, citizens in each of these groups used activism to excite others about the park. Inspiring Memphians from across the city and working to achieve real results, Park Friends and Save Our Shell played an important role in saving Overton Park from prolonged mediocrity.

Founded in 1992, Park Friends was created with one interest in mind—preserving the Greensward. During the Memphis

Zoo's renaissance in the 1980s and 1990s, its popularity grew, as did its demand for on-site parking. Beginning a decades-long battle, the Memphis Zoo expressed an interest in paving a portion of the historic Greensward, which generated significant resistance. Park Friends organized to preserve this beloved green space in the city.

Melanie White, who became a park advocate through running, was part of this group when it formed. "We got a call from the Division of Park Services director saying that he had just got word that the zoo wanted to pave the Greensward and that we needed to put together a group to keep that from happening," explains White. "That's how Park Friends got started." Park Friends was successful in keeping the zoo from covering the Greensward in asphalt, though tension around this area of the park still exists today. But through this one issue, an Overton Park–centric group formed and became a tireless advocate of protecting the park.

Park Friends did not limit itself to the Greensward controversy. This group also led the charge in preventing a senior center from being constructed in the park in the early 1990s, arguing that the park need not be further developed. As the century came to a close, they also organized trash pickup days in the park, helped spread gravel along running trails, and created a system of paths throughout the historic Old Forest. "With a professor from Rhodes College who was on the Park Friends board," explains White, "we developed the

Old Forest trail and put signs in." After the trail had been established, they printed maps of the forest and its trails to encourage Memphians to explore this wooded sanctuary. Their efforts helped visitors become acquainted and comfortable with the forest. Encouraging more and more people to visit the Old Forest, White and her peers reasoned, discouraged some of the "weird activity"—prostitution, public drinking, and blaring music from parked cars—that occasionally went on in the park during the 1980s and 1990s. By doing this work, Park Friends reestablished some of Overton Park's vitality and improved its perception among Memphians.

Park Friends was also integral to maintaining a lengthy Overton Park tradition. In 1999, this group began sponsoring the junior golf tournament. Without their sponsorship, explains White, the tournament would have ended. "It's the oldest junior golf tournament in the country," she says, "and it's played in the park every year in July." Park Friends hated to see this tradition come to a halt. When they got involved, they worked to make the tournament even more welcoming and fun, creating a vision for how they felt it should be operated. "We decided we wanted to make it ten dollars for a whole week of golf." This low price was key. "These kids were not the ones who could afford to play in big tournaments," White explains, and Park Friends wanted to ensure that children could compete regardless of their socioeconomic circumstances. Giving participants custom towels and trophies and feeding them hot dogs, these dedicated volunteers made the junior golf tournament an exciting experience for all involved. Founded on a single issue, Park Friends went on to do decades of important work in preserving, protecting, and improving Overton Park.[12]

Similar to Park Friends, Save Our Shell was a citizens' group founded to protect a specific portion of the park. The Overton Park Shell, built in 1936, was under attack by the 1980s. Rather than maintain the Shell as a venue for concerts and performances, the City of Memphis hoped that the amphitheater would be destroyed and replaced with a parking lot. Like the rest of Overton Park, the Shell had fallen on hard times, and it was no longer the dynamic venue it once was. Its decline prompted some to think that destruction and redevelopment was a better alternative than revitalization.

Fortunately, citizens organized to protect this historic stage. This effort began in the early 1980s with the National Conference of Christians and Jews. Jimmy Ogle, who operated the Shell for a brief four-year period, explains that "there was a movement around the country to honor a Swedish diplomat named Raoul Wallenberg, who had saved [thousands] of Jewish people from Nazi concentration camps during World War II. To commemorate him, we actually renamed the Shell the 'Wallenberg Shell.'" Despite the excitement around the Shell's new name and its role as a tribute to a heroic person, anticipated funding never appeared. The City of Memphis's plan to convert the Shell to a parking lot moved forward. By 1984, the Shell's destruction seemed imminent.[13]

Gigi Wischmeyer remembers that a group of Memphians banded together to stop the Shell's demolition, forming an impromptu organization called "Save Our Shell." Save Our Shell, she explains, "tried to do exactly that.... They just tried to keep [the Shell] from being razed by the City." The organization limped along at first, says Wischmeyer, but they eventually gained momentum and stopped the City of Memphis from destroying this historic venue. In addition to fundraising, part of their preservation plan included offering free shows to encourage popular support for the venue. And even though Save Our Shell sometimes struggled to pay the performers at these events, many musicians offered to play for free because they were honored to perform at this historic venue. "The Shell was on life support, for sure," recalls Jimmy Ogle. But thanks to determined Memphians—especially John Hanrahan, the Hanrahan family, and David Leonard—the Shell survived and was revamped into a beloved Memphis location.

Park-centric citizen groups were instrumental in helping Overton Park push through its lowest points in history. Though these groups operated in different ways, they shared a common theme: they helped Memphians recover a sense of excitement for Overton Park. Citizen groups fostered a renewed sense of stewardship and encouraged Memphians to commit to preserving and enhancing this space. Thanks to these stewards,

Overton Park endured through the end of the twentieth century, preparing itself for a rebirth in the twenty-first.

DIMMER BUT NOT DEAD

Despite the role that runners, park institutions, and park-focused citizen groups played in increasing vibrancy in this space during the 1980s and 1990s, Overton Park still struggled. This strain was most evident away from places such as the Memphis Zoo and the Brooks Museum and in the playgrounds, the Greensward, and the Old Forest. Collectively, Memphians simply were not as excited to frequent these locations with some of the "weird activity," such as people hooking up, washing cars, and dealing drugs, that occurred there. Some citizens nonetheless continued to use these areas within the park. Though not as popular as in the previous decades, the playgrounds continued to attract children and encouraged them to play with friends. The Old Forest beckoned people onto its trails, and the Greensward offered the perfect venue for Recreation Department events, providing fun opportunities for people across the city.

Willy Bearden remained loyal to less popular areas in spite of their suffering reputation. In particular, he recalls taking his children to the playground: "By 1980, my first child was born. So within a couple of years, I really started using the playground a lot because my kids loved to go there." Because he lived so close, Bearden often made the trip to the park without a car. Instead, he equipped his bicycle to carry both himself and his son. "I had a little baby seat on my bike, and we would ride over there," he recalls. Bearden most clearly remembers watching his son on the slide—the same slide that Gigi Wischmeyer and Jimmy Ogle remember as unusually high. "So many of the memories I have of my children [involve the park]," Bearden explains, "like when my son got up the courage to climb the ladder on the big slide. I would sit there, and I'd be so nervous that I'd just have to sit on my hands not to say, 'Well, I'll come help you. I'll come do it.' But I knew that he needed to do it himself."

In addition to using the playground during this era, Bearden and his family also started a tradition of walking through the park on Thanksgiving morning. Since the mid-1990s, he has taken this walk with family and friends each year. "That's what we always do—that is our tradition," he says. "We take the Nerf football, we toss the football around, we take a bunch of pictures, and we just kind of visit. That's what we've done forever." Some of the pictures he took during these walks were of special significance: "I always took a picture of my three children, and that would be the family Christmas card." The red, orange, and golden leaves on the trees made for the perfect backdrop.[14]

Although he did not annually send out photos taken of his family in Overton Park, Jimmy Ogle also enjoyed the space, especially in a professional capacity. As an events supervisor for the Memphis Park Commission Recreation Department, Ogle hosted several events in Overton Park during the 1980s. Two events he hosted during this time—the "Shout It Out" Decathlon and the K-9 Catch & Fetch—stand out as especially exciting occasions for Memphians to come to Overton Park.

The "Shout It Out" Decathlon was a promotional event meant to highlight the effectiveness of Shout laundry detergents. But to show the detergent's power, clothes first had to be dirtied. "So," explains Ogle, "we built ten mud pits near a playground. Organizers gave me this manual on how to build the pits and what obstacles we should put in there." Once the pits were built, children were encouraged to speed through the course and embrace the mud and dirt that was staining their clothes. "Kids would run through the pits to a stopwatch—individuals and tag teams—and at the end of the day, they'd all get a coupon from Shout to get their clothes cleaned," he remembers. The event was wildly successful. But after a few years, Shout stopped sponsoring the event because they had already demonstrated the quality of their product. "But we liked the event so much," says Ogle, "that we called it the Memphis Mud Derby . . . and continued to run this as a playground event." It always took several days to prepare, partly because organizers had to pick through the pits, removing pieces of glass, sharp rocks, and can tabs, to make sure they had "clean dirt"—an oxymoron, laughs Ogle. Even with all the hard work that went into organizing this competition, Ogle and others were always eager to host the Mud Derby in Overton Park.

Given that he had so much success with the Mud Derby, Ogle coordinated another unique event in the park. The K-9 Catch & Fetch was a rotating national event in which dogs would compete in a disc catching contest. "We had the first-ever dog disc contest on the Greensward by Rainbow Lake" in the early 1980s, Ogle remembers. About ten dogs showed up that day, but among these was "the Babe Ruth of dogs, Ashley Whippet." Ashley Whippet, known as the first great disc dog, had participated at Super Bowl pregame events, performed at the White House for President Carter's daughter, and won three world disc-catching championships by the time he came to Overton Park. An American superstar, Ashley brought a wealth of excitement into the park. When planning this event, Ogle could think of no better place to host this canine celebrity and his contemporaries: "Where did we do it? We did it at Overton Park on the Greensward."

CONCLUSION

When the 1980s began, Overton Park was drifting into two decades of difficulty. The park, once an extremely popular destination in Memphis, suffered a knock to its reputation. Rather than viewing Overton Park as the jewel of the city as they once

had, many citizens began to think less of this space. Even in
fiction, the park played host to somber and shady events. In
Peter Taylor's famous short story "The Old Forest," published
in 1985 but set in 1937, the park was the site of a snowy and
scandalous car crash that threatened to ruin an aristocratic
couple's pending marriage. Put simply, the park during this
era is sometimes remembered as a dimmer, duller place than
it was in previous years. Despite its past successes, Overton
Park slowly declined in vibrancy.

But the park didn't die. Thanks to runners, park institutions,
newly formed citizen groups, and loyal Memphians, Overton
Park persevered through the lowest points in its history. Trekking through the Old Forest, along park roads, and around
the golf course, runners gave life to all areas of the park and
helped foster a sense of community among themselves and
with strangers. By reinventing itself, the Memphis Zoo drew
crowds into the park to visit with siamangs, elephants, and
lions. The College of Art and the golf course helped sustain the park through this period, too, inspiring artists with
education and entertaining golfers of all ages, abilities, and
socioeconomic backgrounds. Meanwhile, groups like Park
Friends and Save Our Shell encouraged Memphians to care
about their park and worked to show the importance of this
space to those who tried to damage it. And in the midst of
all these distinct groups and bodies sustaining and reviving

the park, loyal visitors continued to use this place despite its tarnished reputation.

The continued activity in Overton Park throughout these decades prepared this space for a twenty-first-century renaissance. In the 2000s and 2010s, the park, its institutions, and its patrons witnessed a rebirth in popularity and vibrancy. Once again, Memphians fell—and continue to fall—in love with this place. Slowly at first, Overton Park began to be seen again by many as the heart of a now-burgeoning city.

CHAPTER 5

THE HEART OF MEMPHIS

Beginning the Twenty-First Century

Overton Park celebrated its one hundredth anniversary in 2001. To commemorate the occasion, Steve Cohen, then a Tennessee state senator, raised the funds to install a sculpture in front of the Memphis College of Art honoring the milestone. The sculpture—*Ikon* by Ted Rust—was meant to highlight the longevity and importance of this historic public space. Given that Rust was a former director of the College of Art—the director who oversaw the college's move to Overton Park and the school's racial integration—he was the ideal choice to commemorate the park's anniversary through public art. Willy Bearden also celebrated the park's centennial, but in a different way. Bearden, now a famed local filmmaker and historian, made Overton Park the focus of his first hour-long documentary. Taking a year to write and film *Overton Park: A Century of Change,* Bearden wanted the documentary to show that "Overton Park truly reflects everything we are [as a city]." He adds, "It was almost easy to do the film because the park was such a nexus of everything that has happened around here."[1]

Since 2001, Overton Park has undergone a revival, illustrating its significance in twenty-first century Memphis. Beginning with the one hundredth anniversary, Overton Park slowly climbed out of the duller decades of the 1980s and 1990s toward a new level of popularity. This ascent continued with the total renovation of the Overton Park Shell, beginning in 2005. Renamed the Levitt Shell in recognition of the Mortimer and Mimi Levitt Foundation, the outdoor venue began offering fifty free concerts per summer to Memphians in 2008, strengthening the community through music and entertainment. The Old Forest and the Greensward started afresh during this period, too. The Old Forest was designated a Tennessee State Natural Area in 2011, guaranteeing that the forest will be perpetually protected under an agreement with the Tennessee Department of Environment and Conservation. And after months of intense efforts by park stewards and city officials, the parking issues that have plagued the Greensward for decades neared a permanent solution in the summer of 2016. These new improvements began operating alongside decades-long traditions to foster a bustling, vibrant, and hopeful atmosphere throughout the park.

With the park's new improvements and beloved traditions came an increased excitement regarding this shared space. Wanting to ensure that the park maintained its newly redis-covered vitality, citizens and city leaders appointed new park stewards. Groups like Park Friends remained important in the twenty-first century, but a new management organiza-tion—Overton Park Conservancy—formed in 2011 to care for the park, working full-time to sustain and improve this place. Overton Park evolved to again be recognized by many as the heart of Memphis. For well over a century, the park has experienced highs and lows. Like most American institu-tions, its past has included moments that have been dark, dim, and painful. But its present and its future are bright. Given the renaissance it has undergone since 2000, Overton Park is poised to remain a treasure within Memphis for years to come. Its resurgence stands as a tribute to the Memphians who took ownership of Overton Park.

TRADITIONS CONTINUE

Long-standing traditions persisted in Overton Park in the early twenty-first century. Even though the park was chang-ing for the better, not all that was old or entrenched was abandoned. Some of these traditions, such as walking and running through the Old Forest and painting in the park, are considered less formal but more personal. Many citizens have trekked the same path through the forest or painted the same gnarled tree for years. The park, for them, has become a place of habit and comfort. Other traditions—of caring for animals at the zoo, displaying high-caliber art at the Brooks Museum, and organizing golf tournaments for youth—remain more formalized. Affecting vast numbers of Memphians, these established practices help the park hold on to its history even in the present.

Willy Bearden is one of many Memphians who has consistently strolled through the park for years. "I have walked through that park I don't know how many thousands of times," he says. Enjoying the quiet that it offers, the filmmaker is able to relax and think as he meanders through the Old Forest. His route is generally fixed and varies only when necessary: "I usually park at the golf house, and I walk all the way around the road through the Old Forest. But I don't go in the forest during snake season because I've seen too many copperheads over there. They'll crawl when it first starts getting cold in October, and then November and through the winter, you're fine." Blushing, he admits, "I'm scared of snakes." Since copperheads are not slithering through the park in November, Bearden has no fear when he, his family, and his friends take their annual Thanksgiving morning walk. After taking this walk for twenty years, Bearden and his family have come to view their annual stroll as an enduring tradition. "It wouldn't be Thanksgiving morning if we didn't do our walk," he says. "People call me during Thanksgiving week and say, 'Hey, Willy, are we taking the walk Thursday morning?' Some of my ex-in-laws come. It's funny, but it's just a part of who we are."[2]

Willy Bearden took this photo on one of his many walks through Overton Park.

Another person who still enjoys her walks through the Old Forest is Janet Hooks. Hooks, who once led the Parks and Neighborhoods Department for the City of Memphis, prefers to walk alone, which affords her a rare chance to reflect and be fully present. "I like to go solo," she says. "I'm a morning person, and I love walking around the Old Forest on the trail. I like to see how the sun is filtering through the trees. It's quiet. And for me, at that point in time, all is right with the world. There's not a lot of movement, no horns honking. I don't have to be any place at any certain time. I can listen to the birds. I can watch the chipmunks. I can hear the wind as it makes its way through the trees." Hooks finds walking through the forest restorative, and she is grateful to have such a serene and tranquil setting to experience: "For me, it is an issue of renewing my spirit."

She shares this sentiment with others who walk through the forest. "Everyone is friendly, and when you pass people, they nod in acknowledgement. But for me," she adds, "they're nodding in acknowledgment that this is a relaxing environment. It's just an unspoken commonality, if you will. It sometimes is just a nod. But you feel that oneness. You feel as though everyone you see there is taking in the same thing that you are when you're there."[3]

Michael Cody, Sally Jones Heinz, and Gigi Wischmeyer have also maintained their traditions of trekking through the park, but at a quicker pace. Cody is still "in the park running

four days a week." As it did in the 1990s when he started running here, the park offers softer surfaces that make running a
comfortable—rather than painful—experience. "I'm just running twenty miles a week, all on grass or dirt. I cut the mileage
down drastically," he nonchalantly says, as if twenty miles per
week is not an impressive feat for a man in his eighties. With
his knowledge of running, Cody started a running group to
teach others about the trail system in the Old Forest. Like
Bearden, he keeps a close eye out for snakes. He has not had
any problems with the slithering creatures to this point, and
through his class, he has helped to make running in Overton
Park a tradition for others.[4]

Similar to Cody, Sally Jones Heinz often runs through
the Old Forest. "We have a little group that meets at the golf
house at 6:00 a.m. on Tuesdays, Thursdays, and Saturdays,"
she explains. "We have three or four routes, and somebody
will call [the route out] for the day." Through early morning
runs and races held in the park, Heinz has grown familiar with
the Overton Park running community. As it did in the 1980s
and 1990s, the space still facilitates a kinship among runners.
"It's always such a nice community of people who are out,"
Heinz says, and this community is in no way exclusive. "We
had this woman run up to us one time when we were getting
ready to start off from the golf house, and she asked, 'Are you
an organization? Can I run with you?' We were just like, 'Yeah,
come run with us!' She's from Ireland. She had just moved

here. She lives in the neighborhood and works for FedEx, and now she's just a regular with our crew." It is this community that keeps Heinz coming back to the park.

Some, like Gigi Wischmeyer, prefer losing themselves in their runs, allowing their stress to slip away. "I just aim for consistency," she says, explaining that she used to run for exactly one hour, four days per week. Recently, she has cut her duration back to forty minutes. What she loves most about her trips through the park is the opportunity it offers to mentally escape from outside pressures. "I let my mind wander," explains Wischmeyer. "That's why I love being in the park. I'm not on a track or treadmill." It is the process, she says, of running through nature and immersing herself in a place that has inspired her to run in the park for the last few decades.

Runners and walkers are not the only visitors who have maintained informal traditions of using the park. Martha Kelly, a Memphis artist and printmaker, has developed a years-long custom of exploring her artistic passions in the park. "All seasons, all weathers, all years, I paint." For Kelly, there is something comforting in returning to the same place to express herself artistically. "I go back to the same tree," she explains. "I keep doing that tree. I go back to the forest and keep doing the forest." When she goes on her painting trips, she sometimes carries her banjo so she can make audible art, and she also takes her dog, Mr. Darcy. "We set up, and he hangs out with me while I paint," Kelly smiles. And when

A linoleum block print by Martha Kelly
of her favorite tree in the park.

Kelly decided to develop her artistic skills even further, she
returned to the College of Art and took a few classes as an
adult. "I was ready to do something different than oils, and
I took a printmaking survey class and fell in love with relief
printing. That's been huge for my career," she says. Twice now,
Kelly has fallen in love with art in Overton Park. Given her
passion for painting and sketching park scenes and landscapes,
she plans to continue painting, learning, and falling in love
with art for years to come.

Such informal traditions contribute to the eclectic nature
of Overton Park in the twenty-first century. To know that
some citizens have continued to use the park in the same

capacity for decades creates a sense of permanence. More institutionalized traditions are essential to the park's durability as well, perhaps even more so than those of individuals. The Memphis Zoo, the Brooks Museum of Art, and the Links at Overton Park have all played key roles in maintaining many Memphians' favorite aspects of Overton Park. By committing themselves to build upon their past and always strive for excellence, these institutions have ensured that tradition is not defined by boredom or stagnation.

After reinventing itself in the 1980s and 1990s, the Memphis Zoo underwent a boom in popularity. Now ranked as one of the top zoos in the country, the Memphis Zoo is a favorite destination for both locals and tourists. According to James Jalenak, when the Zoological Society took over management from the City of Memphis in the mid-1990s, the zoo attracted about 550,000 visitors per year. Those numbers are now very different, Jalenak explains: "Our attendance has been over one million people for the last three years."[5] These visitor statistics alone reflect the Memphis Zoo's reputation across the city and region.

The zoo transformed into such a popular institution, in part, because it initiated standards of world-class animal care, high-quality education, and commitment to improvement. Kathy Fay and Richard Meek, both retired zookeepers, attest to the care that staff provide for the animals in the zoo. "Right now, the level of professionalism there is just breathtaking," says Meek.

But just because the keepers and caretakers at the zoo take pride in their professionalism does not mean that they have abandoned principles of compassion, kindness, and love. Meek remembers several stories of the strong relationships between animals and their keepers. One of his favorites is of a giraffe named Angela Kate. "I bottle-raised her because her mother rejected her," he explains. "She considered us her parents." Smiling, Meek says that Angela Kate "would follow me around like a little puppy dog." After raising her from infancy, Meek felt a particular bond with the giraffe. But Angela Kate was not the only giraffe that he worked with. For about ten years, Meek worked full-time with these gentle giants, caring for them on a daily basis. "I had the best thing with giraffes. Honestly, I did," he laughs. He particularly loved exposing zoo visitors to what he considers the sweetest, most gentle creatures on earth: "For years, until they opened the giraffe feeding platform, we used to do behind-the-scenes tours of giraffes for people. I would do it for just anybody that asked, but supposedly the zoo was supposed to arrange it. Usually, they did it for people with a lot of money. So I would take people—families, little children—back into the barn with the giraffes. These kids are standing there, and there's a seventeen-foot-tall giraffe standing over them. I'm telling them all about the giraffes, and the giraffe leans down to take a piece of banana from them. This seventeen-inch-long tongue comes out and wraps around them. And I got to watch that almost

Richard Meek bottle-feeding Angela Kate the giraffe.
Photo courtesy of Richard Meek.

every day, and it was wonderful. I got to watch the children just go crazy."

Like Meek, Fay also experienced a special bond with one of the animals she looked after. "Without a doubt," says Fay, "[my favorite animal] was a particular sea lion named Chloe. I have her picture in my wallet." Fay became close to Chloe for many reasons, especially because of the vast amount of time she spent training her: "I was probably spending more time and paying more attention to [Chloe] and my other zoo animals than I was to my dogs and cats at home. You get close to your animals, and you fall in love with them."

Fay did not work solely with sea lions. She worked with pandas for a few years and trained elephants, too. She also had the dream job of many animal lovers, taking care of baby animals at the zoo. "I did a lot of hand-raising of babies," she explains. "I raised some really cool stuff, like orangutans and snow leopards. And bats, I loved hand-raising the bats." Her care for the baby animals sometimes went beyond the gates

Kathy Fay teaching children about sea lions at the Memphis Zoo. Photo courtesy of Kathy Fay.

Kathy Fay taking care of baby animals, a dream come true
for a lifelong animal lover. Photo courtesy of Kathy Fay.

of the zoo. "Frequently, the babies would get carried back and
forth because they needed twenty-four-hour care, so I would
carry them home." She loved taking care of the animals, but
her own pets were troubled when Fay brought her work home.
Whenever she carried baby animals to her house, she had to
separate her dogs and cats from having any contact with the
animals from the zoo. "They weren't particularly happy about
that," she laughs. Her dogs and cats were especially displeased
when Fay was raising the orangutan. "He was pretty disruptive in that he needed to be with me all the time. So basically,

the cats were elsewhere and ignored until he grew up." With what was arguably the cutest job at the zoo—but one that was demanding and time-consuming—Fay gained an extremely rare experience in becoming the primary caretaker for several baby animals.

In the same way that the Memphis Zoo maintains traditions of world-class animal care, it also continues to provide opportunities for education. Again, Fay and Meek offer the best insights into how the zoo educates its visitors. They recall doing "keeper chats" daily to teach the public about the animals. "At a specific time of the day, we would go out and talk to the public—whoever happened to show up—about our animals," explains Fay. "So for a lot of years, I talked about pandas every day at 11:00 a.m., and [Richard] talked about giraffes at 1:55 p.m." The length of time they spoke often varied: "You could talk anywhere between five minutes and twenty minutes, depending upon how many people were there. . . . Those interactions were some of the magic ones where the kids could get really excited and you could show them things. You could talk to them."

Sometimes, Fay's and Meek's interactions with visitors were not limited to specific time slots. They would often educate even as they cleaned out exhibits. "We would clean exhibits right there in front of everybody, of course, because there's not enough time to get it all done before the public gets there," says Fay. Meek explains that while they cleaned exhibits, keepers often interacted across the fences with the public. "It usually started with, 'What's supposed to be in here?' or 'Where are the animals?' But it could then progress to really, really fun stuff and great educational opportunities."

Meek notes that often children and even adults are having so much fun at the zoo that they do not realize they are learning new things. "They just absorb it," he says, "and they don't even know what's happening to them. It's wonderful. It's sneaky education at its best." One of the best examples of "sneaky education" can be seen through Fay's sea lion show. They did shows that taught about the physiology and biology of sea lions, she says, but one of the most popular routines highlighted the importance of recycling. "The sea lion would get the bottle out of the water," Fay says, "and put it in a recycle bin." This routine, unbeknownst at first to Fay, inspired children to recycle and reduce pollution. "One of my fondest memories is when someone sent me a letter describing an interaction between two of her grandchildren," she recalls with a smile. "She was listening to a whole bunch of hubbub in the bedroom, so she peeked in the door, and one was on the bed and one was on the floor. They were playing 'sea lion show.' And one was me, and the other one was the sea lion. One was jumping off the bed and getting the bottle so they could put it in the recycle bin. That's one of my fondest memories. I've got that letter in a safe place." Because of Fay and her sea lion shows, those two children—and countless others—had recycling engrained in their minds. The sea lions' routine affirms Meek's belief that, at the zoo, "you can educate at such a high level without the public even knowing they're being educated."[6]

The final tradition that the Memphis Zoo has held strongly to throughout its history—especially its recent history—is adding new exhibits and improving infrastructure. As in the 1990s with Cat Country and Primate Canyon, the zoo added an entirely new facility for its hippos in 2016: the Zambezi River Hippo Camp. While this latest addition is the end of a long plan to improve the zoo, James Jalenak knows that the zoo will soon push itself further to continue increasing its quality. "Zambezi is the end of our master plan," he says, "so we're in the process of starting a new master plan. That'll probably take a year or so to get done. There are a lot of improvements needed, especially in the west end of the zoo." Given this record of striving for improvement, the Memphis Zoo will continue to bring local Memphians and tourists into Overton Park.[7]

Like the Memphis Zoo, the Brooks Museum of Art is a popular destination within the park and has been integral in its renaissance since 2001. It has maintained a tradition of displaying superb pieces and attracts art lovers of all kinds. Sally Jones Heinz, who worked at the Brooks Museum for a few years in the early 2000s, describes the Brooks as "a little jewel sitting there in the middle of the park." Its architecture

Children and families making chalk art outside the Brooks
Museum. Photo credit: Melissa McMasters, director of
communications for Overton Park Conservancy.

and its gallery are beautiful, she says, and its location in the
center of the park is crucial. "To be able to pair art with nature
and have that be in sync—it's really nice. It's just fabulous to
have a museum situated in a green space like that."[8]

Martha Kelly also admires the Brooks Museum. As an artist, she often finds herself painting the outside of this beautiful

building. "I probably paint the outside of the Brooks more than I go into it at this point," she admits. But she appreciates the wide variety of prints and paintings housed in the museum. She has been a member of the Brooks for years and loves to show the museum to her friends from out of town. One of her favorite memories occurred in 2001 when pieces from the Scottish National Gallery were housed in the Memphis museum. "The Scottish National Gallery in Edinburgh is one of my favorite museums in the world," she explains. "It has Constable, who is my favorite painter, and they have one of my favorite Constable paintings. [The Scottish National Gallery] was being renovated, and their entire collection went on tour—the stuff that doesn't normally go on tour. The Constable that I love was in the Brooks for a summer." Kelly took advantage of her favorite painting being right across the street. "I went over twice a week and just sat in front of it," she remembers.[9]

The golf course in Overton Park has held tightly to tradition as well, especially in hosting the oldest junior golf tournament in the country. Since 1945, the golf course has hosted countless children who competed in this tournament. Even severe storms, remembers Melanie White, could not stop this tournament from being held each year. In July 2003, a storm locally known as Hurricane Elvis passed through Memphis in the early morning, bringing winds in excess of one hundred miles per hour. "It took all the power out, all the trees out, and

we had no electricity for a week," remembers White. "All of
this happened on day two of the junior open." Initially, she and
other Park Friends tournament organizers were unsure if the
competition would continue. They began the tournament that
morning but soon had to postpone. White recalls that young
players continuously called in to ask about the tournament's
status: "They kept calling in and asking, 'Do we have a tee
time?' We said, 'Look, there's no power. It's hot. You can't go
into the bathrooms because they're pitch black.'" But after re-
ceiving so many phone calls, White and her co-organizers got
together and decided that, despite the obstacles, they would
finish the tournament, shortening the rounds but continuing
with the competition.

They made the right choice. "It was most fun," explains
White, "because a lot of people weren't working because there
was no power in the city. So parents came to the junior tournament [to cheer on the children]." Though it took an entire
day to get the course clear enough to resume, a festive atmosphere characterized the final two days of competition. "People
came and set up big barbecues on the clubhouse porch and
fed the kids. It was such a neat community feeling," White
recalls. "That was just wonderful." White and other tournament organizers could have easily canceled the junior open in
2003, and no one would have blamed them. Hurricane Elvis
was, after all, one of the strongest storms that Memphis has
weathered. But because they were creative and committed,

the golf course in Overton Park continued its tradition as the oldest annual junior golf tournament in the country. Each year, young golfers are delighted to compete on the Overton Park course, creating a sense of energy, excitement, and happiness in the park during the tournament.[10]

Whether through personal or institutional traditions, consistency has remained an important part of the Overton Park experience, creating permanence and stability within the park and maintaining a fluidity between new happenings within this place. Park users and institutions respect the history of the park and hope to continue this history, but they do so without threatening progress and improvement, an important distinction for an ever-evolving place within the city.

NEW BEGINNINGS FOR THE SHELL, THE OLD FOREST, AND THE GREENSWARD

New beginnings in the park have played a role as important as long-standing traditions. Although the park has, in some ways, reinvented itself, three new beginnings stand out as particularly important. When the Overton Park Shell became the Levitt Shell in 2005, this facility started its climb back into the Memphis spotlight. In 2008, when it officially reopened after a few years of renovations, park visitors were amazed by the energy at the Shell and thrilled to have this entertainment option right in the middle of the park. The

Old Forest experienced a rebirth during this era, too. In 2011, thanks to the Tennessee Department of Environment and Conservation, the Old Forest became a State Natural Area, a designation that protects it from all manner of encroachments, ranging from development and vehicular traffic to the removal of plants by individuals. Finally, the Greensward in Overton Park recently began its life anew. For decades, this historic green space has been used for overflow parking by the Memphis Zoo. After several months of heated discussions involving the mayor, the City Council, the zoo, Overton Park Conservancy, and other park-oriented groups, a compromise was finally reached. In July 2016, the parties entered into an agreement to permanently end parking on this grassy area within the next few years.

The Levitt Shell's impact is most clearly seen after its free concert series began in 2008. Behind-the-scenes work, however, was crucial in getting the Shell to this point. Without the work that began in 2005, the Shell would never have experienced a rebirth. Blanchard Tual, who served as president of the Shell in 2012 and 2013, was part of this effort to reenergize the venue before its grand reopening. "In 2005," he says, "Thomas Boggs and Barry Lichterman got together with another small group of people, and I was one of those people. We formed a committee to try to renovate the Shell, and we partnered with the Levitt Pavilions. They provided us with some seed money to renovate the Shell, along with some money we got from the City of Memphis." This financial support by the Levitt Pavilions—an organization whose mission is to reinvigorate America's public spaces through creative place-making and create opportunities for everyone to experience the performing arts—was not only important in revamping the Shell, but remains essential today: "The Levitt Pavilions still provide money for us to operate each year. We're very thankful to them," he says. Since his early involvement with the Levitt Shell, Tual—who still attends an estimated 90 percent of Shell concerts—has played a key part in nurturing the venue's success.[11]

Anne Pitts—the executive director of the Levitt Shell since 2008—offers a glimpse into the excitement that it has brought to Overton Park. Pitts, who practiced intellectual property law and entertainment law for a number of years, moved to Memphis in 2008 to take up her new job at the Shell. She remembers that this position was, in the beginning, rather difficult. The idea of hosting fifty free concerts per year, as the Levitt Shell planned to do and still does, was "very much pie in the sky at that time," she says. "I knew right away that my first responsibility was to convince the community that this could happen, that this idea was for real." Everyone in the community, especially those who helped renovate the Shell, was very supportive, she remembers. "But everyone was still wondering, 'How do we make this successful?'"

To convince others that this model of fifty free concerts could work and help both the Shell and the park thrive, Pitts embraced public relations as her first role as executive director. "I was hitting the pavement and going to person after person, place after place, and convincing everyone that this was a sustainable model. Because the thought was, 'You're going to take this old building and present fifty concerts, and you're going to make them all free? And you're going to start this right when the market has crashed?'" Pitts faced an uphill struggle.

But putting so much effort into public relations paid off. Within two years, Memphians began to believe again in the idea of the Levitt Shell, and once confidence grew across the city, Pitts and her crew started working to make the Shell a place that strengthens community. "The mission of the Shell— the reason we present fifty free concerts—is because we want to build a stronger, more connected community in Memphis," she says. "What better place to do that than a public park where everyone is welcome? And what better way to do it in Memphis, Tennessee, than with music?"

Offering diverse entertainment options was one way the Shell tried to nourish community values. Instead of following other models that present specific genres on certain nights, as some free concert venues do, Pitts and her team diversified the artists who performed at the Shell and when they performed. When looking at those other models on paper, she explains, it gives the sense that "this night is 'white night,' this night is 'Latino night,' this night is 'black night.' And that's not what

Memphis music is about." Noting that "people here love all kinds of music," she says the Levitt Shell made an intentional effort to eliminate genre-specific nights. At the Shell, she explains, "You can go on a Friday night, and who knows what you're going to hear? It could be jazz, it could be Latin, it could be a band from Nigeria—who knows? It's the art of discovery. It's the joy of discovery. I think that's one of the things that has made the Shell so successful." Presenting such a variety of music on different nights encourages Memphians from all backgrounds and neighborhoods to come to the Shell and enjoy music. Being in this atmosphere together, Pitts believes, fosters positive feelings toward others and helps individuals care more for one another. The sense of community the Shell fosters is, Pitts says, its most important feature.

Amid all of the hard work she has put into reviving the Shell, Pitts has experienced some magical moments in her role as executive director. Hearing performers talk about their experiences at the Shell is one of her favorite parts of the job. "What I love is when the artists come during the concert season, and they're looking backstage at the different photographs that we have of Elvis Presley and Carl Perkins and the paintings that we have of Alex Chilton and Phineas Newborn." Because such artists and countless other legends of music have performed on the concrete stage, the Shell is considered "sacred ground." When performers step into the spotlights, they immediately feel a connection with their forebears. For most artists that come through, says Pitts, at least "one of their heroes has been on that stage, which is pretty powerful stuff." Knowing that the established legends and future stars have played on this stage in Overton Park makes for an energetic entertainment atmosphere.

An example of this connection to the past is illustrated through recent performances by Lisa Marie Presley and Roseanne Cash. "When people think of the legends of music who have played the Shell over the years," Pitts says, "they obviously think of Elvis Presley and Johnny Cash. In the last

Lisa Marie Presley performs with her band on the Levitt Shell stage. Photo by Andrea Zucker; courtesy of Levitt Shell.

Rosanne Cash playing to a capacity crowd at the Shell.
Photo by Andrea Zucker; courtesy of Levitt Shell.

few years, we've paid homage to that history. [In 2013,] we had Lisa Marie Presley perform, and then [in 2014,] we had Roseanne Cash perform. And it was amazing. Their experiences of standing on the same stage where their fathers got their starts—that was a pretty powerful experience. I'm sure it was powerful for them, but for me personally, it was just incredible to see that history come full circle."

Yet another performance stands as most memorable to Pitts: "the first concert as the Levitt Shell—September 4, 2008." Not only did she enjoy the entertainment that Jim

Dickinson and Amy LaVere provided, but it was at this show, Pitts says, that Memphians began to believe in the Shell. For the months preceding this event, Pitts had been promising people that, despite their doubts, the Shell would be successful. In an effort to revitalize the venue, she and her team had made some changes to the Shell, including painting over the iconic rainbow that had long covered the back wall of the stage. As Pitts feared, these changes made people hesitant. But in moving forward, the new stewards of the Shell did not entirely abandon the building's history. Putting in LED lights into the coves of the Shell—strategically designed in the colors of the rainbow—connected the Shell to its past while heralding a brighter future. "The lights came on, and they made a rainbow. You could just hear the gasps through the audience. It was such an exciting moment to see the delight on people's faces, and all of a sudden, it was like people really saw how magical the Levitt Shell could be."[12]

With the revival that started in 2005, gained momentum in 2008, and continues today, the Levitt Shell has made a tremendous impact on Overton Park's vibrancy. Willy Bearden says that the Shell has "completely revolutionized the community's thoughts of Overton Park. That one thing—the Shell—has really done more than anybody realizes now. It has brought people together, and it has invigorated that place. It has given us a very different way to view an outdoor concert venue."[13] Blanchard Tual suggests that the Shell has also improved race relations in Memphis. "When people walk in that area," Tual says, "they almost feel like they're in a church. It's almost reverential to people."[14] Since 2008, the Levitt Shell has done much more than just make music. It has helped create a community within Overton Park.

The next area to begin anew in Overton Park was the Old Forest. In June 2011, the Old Forest in Overton Park was designated a State Natural Area by the Tennessee Department of Environment and Conservation, guaranteeing its perpetual protection. As with many decisions made in Overton Park, the designation of the Old Forest did not come without controversy. Charlie Newman, who helped save the Old Forest from destruction in the 1971 Supreme Court case *Citizens to Preserve Overton Park v. Volpe,* says that the forest was in danger because of the Memphis Zoo. Because he views the forest as "the most irreplaceable asset in the park," Newman was concerned by the zoo's gradual encroachment. "The zoo had begun to take small pieces of the forest, getting the City Council to give them about seventeen acres" for a new exhibit that was built in the mid-2000s. Much to the outrage of Park Friends and the recently reincorporated Citizens to Preserve Overton Park, the zoo had clear-cut a few acres to construct Teton Trek, an exhibit that pays homage to the Pacific Northwest.[15] So when the rest of the forest went up for protection, explains Newman, the zoo expressed disapproval of the State Natural Area designation.[16]

Ultimately, despite some objections by the zoo and by citizens who mistakenly feared that the new designation would prevent them from running or biking through the forest, the Old Forest was protected. "We're glad," says Shelby County historian Jimmy Ogle, "that we have 126 acres of Old Forest [preserved in Overton Park]."[17] When many walk through this permanently preserved place, they immerse themselves fully in nature. Since retiring, former zookeepers Kathy Fay and Richard Meek walk through the Old Forest every day of the year, often noting how the forest changes from season to season. "Some of the most dramatic changes can be seen when you're on some of the interior trails. In the wintertime, it's so wide open. Of course, any time there's a snowfall, it's magical. And it's fun to watch in the spring as the leaves come out. . . . [The leaves] cut the noise out," so you can hear nothing but the forest when traipsing through. Walking through the Old Forest after it became a State Natural Area, park visitors can let their minds wander. They can delight in the changing seasons. They can enjoy a conversation with a friend or a stranger. One thing they do not have to do is worry that the Old Forest will be damaged.[18] The Tennessee Department of Environment and Conservation eased that concern in June 2011. Tennessee's state government, which once worked tirelessly to mow down the forest, now protects it.

The final area in the park that has only recently begun to shift toward a new beginning is the Greensward. A compo-

A runner enjoying the trails through the Old Forest.
Photo credit: Melissa McMasters, director of
communications for Overton Park Conservancy.

nent of George Kessler's original design of Overton Park, the
Greensward has been an area of tension for the past several
decades. Ever since the zoo began using the Greensward for
overflow parking on its busiest days, there has been opposition
from some citizens, government officials, and park-oriented
groups. The tension between the zoo and these parties grew

exponentially when, on December 31, 2015, City Council Attorney Allan Wade issued a legal opinion stating that the Memphis Zoo, rather than Overton Park Conservancy, controlled the Greensward. In reaction to this statement, protesters took to the park to try to stop cars from parking on the Greensward, and they continued to do so for months.

Citizens and officials differed on what they believed to be the appropriate course of action within Overton Park. James Jalenak, a former official at the zoo, explains that the zoo was very upset by the actions of protesters who tried to stop the parking on the Greensward. The zoo's leaders felt that they were doing nothing wrong. On the other hand, countless park supporters criticized the zoo for continuously damaging what is the largest expanse of green space in the interior of the city. They wanted this space to be used for its original purpose.[19]

All the debate surrounding the Greensward is thankfully leading toward a compromise. After decades of disagreement between park supporters and the zoo, Mayor Jim Strickland issued a resolution in July 2016. The resolution stated that the Memphis Zoo will gain control of a sliver of the Greensward adjacent to their current parking lot, while Overton Park Conservancy will continue to manage most of this space. Additionally, the zoo will reconfigure their parking lot and add nearby street parking in accordance with a conservancy-funded spring 2016 study, actions that should add several hundred spaces. When this is done, the Greensward will cease to be used as a

Memphians (and their dogs) gathering on the Greensward near Rainbow Lake. Photo credit: Melissa McMasters, director of communications for Overton Park Conservancy.

parking lot and will again be home to romping children and Frisbee throwers. Although the final details of the plan are still under discussion and could evolve, the Greensward appears destined to begin anew within the next few years.

New beginnings—whether for the Levitt Shell, the Old Forest, or the Greensward—have been crucial to Overton

Park's renaissance in the twenty-first century. In conjunction with its long-standing traditions, these recently resurgent areas boosted Overton Park's reputation among Memphians and helped this place become the heart of the city.

THE STEWARDS OF OVERTON PARK

With long-standing traditions and new beginnings strengthening Overton Park since 2001, the park has reached new heights of vibrancy. But as its history shows, Overton Park has been threatened even during periods of prosperity. Because of this fact, it was important for Memphians to act as stewards of the park and ensure that this newly rediscovered vitality was not lost again. Two main groups—Park Friends and Overton Park Conservancy—have been responsible for the park's wellbeing since the beginning of the twenty-first century, while other smaller groups have also cared for the park. Though all of these organizations operate independently, they have each been crucial to Overton Park's recent success.

In the 2000s, Park Friends fulfilled the roles of both park advocate and park protector. Memphis artist Martha Kelly has been an active member of Park Friends for about ten years now, and she has even served as president of the organization. "We have done a lot of volunteer work," she says, reflecting on her time with the group, "and we used to do a lot of advocacy." Kelly explains that throughout the group's history, the City of Memphis has either led or supported projects that sought to develop parts of the park. Park Friends helped stop a cell phone tower from being built in the park in the early 2000s, and they later prevented the Greensward from becoming a storm-water retention basin. "The City decided they were going to dig up the Greensward," she elaborates. "I spent a huge amount of time over a two-and-a-half-year span doing park advocacy, reading storm-water engineering reports on Saturday nights, doing all kinds of stuff." When Kelly and Park Friends finally succeeded in stopping the city from digging up the Greensward, city engineers helped ease the tension between the opposing parties. "They brought this cute little certificate when they officially killed the plan to build [the basin] in the park. The certificate read, 'The Greensward Plan is Dead,'" she remembers.

Kelly recalls another memorable effort led by Park Friends, one that aimed to improve the quality of Rainbow Lake. Though the park was not threatened in this scenario as it was in others, the project to clean the lake was still important, and it shows the benefit of cooperation between park institutions and park advocacy groups. In the summer of 2009, the City of Memphis planned to drain Rainbow Lake to clean the clogged fountains. "A whole bunch of people came to Park Friends and said, 'Could you do something about the fish?'" Kelly explains. "And we had Brian Carter on our board at that time, sitting in for the zoo, and he said, 'We've got some empty aquariums in the reptile center.'" Working with five-gallon buckets, golf carts, and fish nets, Park Friends volunteers and Memphis Zoo employees worked to save the fish in Rainbow Lake. "We saved seventeen turtles and several hundred fish. We spent two days wading in Rainbow Lake, netting up fish," Kelly says. "The zoo workers were driving back and forth, dumping in the aquariums, and coming back out. It was a massive amount of work." To celebrate the successful cleaning of the lake, Kelly and the other workers engaged in some friendly competition. "When we got ready to release the turtles back into the lake, we all took a turtle, and we had a race to see whose turtle would make it to the lake first."[20]

In many different ways, Park Friends has acted as a faithful steward of Overton Park in the twenty-first century. Other groups have helped care for the park during this time, too, and their efforts must not be underappreciated. In 2008, a group of Memphians formed a new version of Citizens to Preserve Overton Park. This group, which is unaffiliated with the original organization that stopped the interstate in the 1970s, aimed to increase support for the Old Forest and the Greensward. They have led walks through the forest to educate Memphians about the area's ecology, and they voiced strong disapproval of the zoo's 2008 clear-cutting of an area of the forest to install a new exhibit. This Citizens to Preserve Overton Park group reenergized in the 2010s to oppose the Memphis Zoo's parking on the Greensward. Starting a campaign called "Save the

Greensward," CPOP placed banners and yard signs across the city and garnered widespread support from Memphians. They also protested the parking in person, proudly wearing their bright green shirts, and they offered alternative parking solutions along neighborhood streets to interested zoo visitors.

Like CPOP, the local chapter of the Daughters of the American Revolution also took responsibility for a specific area of the park. Melanie White, a member of this group, explains that the DAR has "kind of adopted the formal gardens." Pulling weeds and planting flowers, these women help maintain one of Overton Park's most beautiful areas. As evidenced, citizen advocacy groups have been essential to sustaining Overton Park in the twenty-first century.[21]

Despite the impact that groups like Park Friends, Citizens to Preserve Overton Park, and the Daughters of the American Revolution have had on the park, Overton Park Conservancy has done the most of any organization to maintain and improve the park in recent years. As a nonprofit with a staff dedicated solely to caring for the park, Overton Park Conservancy has helped the park become an even more prominent destination within the city. Charlie Newman, who has done pro bono legal work for the conservancy and helped negotiate the management agreement between the City of Memphis and the conservancy, remembers the beginnings of the organization. In 2011, he explains, George Cates, a prominent philanthropist in Memphis, played a tremendous role in creating the

Staff members from Overton Park Conservancy greeting bicyclists as they enter the park. Photo credit: Melissa McMasters, director of communications for Overton Park Conservancy.

conservancy. Tirelessly, Cates recruited dozens of Memphians to support the organization and helped raise the necessary money to get the conservancy on its feet. Since this time, Overton Park Conservancy has been the primary steward of the park. Others, including the conservancy's current executive director Tina Sullivan, have picked up on Cates's initial mission and have helped the park make great strides. According to Newman, "The park has really prospered under the conservancy's leadership."[22]

Thanks to the conservancy's remarkable efforts, Overton Park has improved immensely. The conservancy has revamped playgrounds, built a dog park named "Overton Bark," organized festivals, and maintained the Old Forest while also creating community partnerships, raising millions of dollars, and managing the park's daily operations. "The conservancy is putting its money where its mouth is," says former Levitt Shell president Blanchard Tual. "They're committed to making the park's quality a first-class priority." Willy Bearden echoes Tual: "I was on the board of Park Friends, and I think we've effected a lot of change. But I'm so glad the conservancy is there now because I never felt like I had the time to give the park all it needed. There needs to be somebody full-time looking after the citizens' best interests in the park, so I'm glad to see the conservancy doing that now." Since its founding in 2011, Overton Park Conservancy has markedly increased the park's vibrancy and will continue to do so until at least 2021, when their management agreement with the City of Memphis is up for review. With selflessness and resolve, Overton Park Conservancy has ensured that future generations of Memphians will continue to enjoy Overton Park.[23]

Much has changed in Overton Park since this book was first submitted. In some ways, this is not surprising. A place as popular as Overton Park is constantly evolving in light of changing circumstances. In other ways, the change is shocking, even potentially worrisome. Although the Brooks Museum has been a park fixture for more than a hundred years, its leaders have announced that they are considering a move to the riverfront. Plans are not yet finalized as of this writing, but even the consideration of a move is unsettling. It is hard to imagine Overton Park without the Brooks Museum.

Another hallmark institution has already shared its intention to cease operations. In October 2017, the Memphis College of Art announced that after fulfilling obligations to currently enrolled students, their doors will close. In a public statement, MCA explained that "the Board of Directors of the College, facing declining enrollment, overwhelming real estate debt, and no viable long-term plan for financial sustainability, has voted to stop recruiting new students, effective immediately, and begin making plans to close the College." It is expected that the institution will remain open through May 2020 to give current students the chance to graduate.

These changes and others, including an intense storm in the summer of 2017 that toppled dozens of trees, will have a lasting effect on Overton Park. They also arouse some uncertainty about the park's future, and beyond uncertainty, they stir concern. If both the Brooks Museum and MCA are gone, what will fill these spaces? What will happen to the park?

Those with questions about the future of Overton Park would do well to examine its past. As this book has shown, throughout its history Overton Park has experienced significant change. Despite—or, perhaps more accurately, because of—this change, the park has flourished. Few places that have been around for over a century remain static. Like the seasons that alter the park's Old Forest, change comes and goes, and yet the forest still stands, inviting visitors to come enjoy its orange and yellow leaves, its snow-covered ground, or its lush undergrowth. While a new season brings visible change, it also brings opportunities for growth.

History has proven that because of dedicated Memphians, Overton Park can persevere and flourish. Many will be sad to see the Memphis College of Art close, and if plans to relocate the Brooks are finalized and the museum's collection heads downtown, park patrons will miss the museum. But Overton Park will continue to evolve in unexpected ways. Opportunities for progress—opportunities that include the eradication of parking on the Greensward, the possible revitalization of an adjacent city maintenance facility into parkland, and a new, highly anticipated master plan for the park—will spur the park forward. Indeed, on the eve of the fiftieth anniversary of *Citizens to Preserve Overton Park v. Volpe,* which helped stop a change that *would* have destroyed the park, Memphians can find grounds for hope. As long as we continue to nurture affection for this place, it will thrive.

Overton Park has been an important feature of Memphis for over one hundred years. But more so than ever before, the park has recently emerged again as the heart of Memphis, a place where citizens can explore art, see majestic lions, and hear Grammy-winning musicians. They can learn how to paint and play golf. They can run through the Old Forest and take their kids to the playgrounds. The park's variety contributes to its vibrancy. "As I look back," says Willy Bearden, "Overton Park has served me in every facet of my life. What more is a park than somewhere to serve you, a citizen, in things that you have an interest in?" Bearden speaks for many when he describes Overton Park in this way, and his question, though rhetorical, is a powerful one. A dynamic place, Overton Park offers something for all visitors.[1]

It is because of the impact that Overton Park has had on the lives of its visitors that many Memphians call this place the heart of the city. "[Overton Park] has such variety and such personality and such longevity, and it's right in the heart of the city," says Jimmy Ogle. "It is just an unusual place. It is the heartbeat of Memphis." Ogle is not the only person to feel this way. Janet Hooks believes that "Overton Park is a jewel in the heart of the city," and Richard Meek describes it as "the most magical park I know of in the country. And as far as I'm concerned, it is the gem of Memphis. It's the best thing Memphis has."[2]

Throughout its history, Overton Park has been the park of the people. It is where Ernest Kelly cultivated a relationship with his grandfather, where a small group of activists joined to fight an interstate that threatened to destroy the park, and where Johnnie Turner determined that she would fight discrimination and racial inequality for the rest of her life, even inspiring her work as a state representative today. Overton Park still belongs to Memphians. New generations are making memories in this place, just as their forebears have done. It remains, and will continue to remain, the people's park. "That's the thing I find really great about Overton Park," says Willy Bearden. "It's ours."[3]

Chapter 1

1. George Kessler, 1911, quoted in William Bearden, *Images of America: Overton Park* (Charleston, S.C.: Arcadia, 2004), 15.

2. Many institutions in the park have experienced name changes throughout their history. The Brooks Memorial Art Gallery is now known as the Brooks Museum of Art. Further, the Overton Park Shell, now known as the Levitt Shell, has undergone several name changes: the Shell has also been referred to as the Memphis Open Air Theatre (MOAT) and the Raoul Wallenberg Shell. The Memphis Academy of Art has also changed names since its founding. It is now known as the Memphis College of Art.

3. George F. Chadwick, *The Park and the Town: Public Landscape in the 19th and 20th Centuries* (New York: Frederick A. Praeger, 1966), 183.

4. Frederick Law Olmsted, "Public Parks and the Enlargement of Towns," 1870, quoted in Alan Trachtenberg, *The Incorporation of America: Culture and Society in the Gilded Age* (New York: Hill and Wang, 1982), 109–10.

5. Bearden, *Images of America: Overton Park,* 13.

6. Trachtenberg, *The Incorporation of America,* 109.

7. In her insightful book *The Politics of Park Design,* Galen Cranz draws a strict distinction between city and country life. The country, she says, stands for simplicity, peace, and stability,

while the city is seen as "too big, too built up, too crowded, diseased, polluted, artificial, overly commercial, corrupting, and stressful." Galen Cranz, *The Politics of Park Design: A History of Urban Parks in America* (Cambridge, Mass.: MIT Press, 1989), 3.

8. Ibid., 42.

9. Theodore J. Smergalski, "The Relation of Supervision to Play and Recreation," *Parks and Recreation* 1 (July 1918): 13.

10. Bearden, *Images of America: Overton Park,* 127. Again, it is vital to note that for decades, only white Memphians were allowed to play on the Overton Park golf course.

11. "Memphis Playgrounds Lead Other Cities," *Memphis Commercial Appeal,* July 18, 1915. At this time, only white children were allowed to use the Overton Park playground. Because racial oppression and ethnic superiority were taught to these children, even if subconsciously, we must be wary of the "ethical behavior" instilled in the youth.

12. Bearden, *Images of America: Overton Park,* 22–23.

13. John Linn Hopkins, "Overton Park: The Evolution of a Park Space," Sept. 1, 1987. This brief history of the park was prepared for Ritchie Smith Associates as they created a new master plan for Overton Park in 1987.

14. Cranz, *Politics of Park Design,* 101.

15. Ibid.

16. Bearden, *Images of America: Overton Park,* 92.

17. Michael Cody, interview by Brooks Lamb, July 2, 2015.

18. Charlie Newman, interview by Brooks Lamb, June 17, 2015; Steve Cohen, interview by Brooks Lamb, September 3, 2015.

19. Cohen interview; Cody interview.

20. Cody interview.

21. Ibid.

22. Ernest Kelly, interview by Brooks Lamb, June 30, 2015;
    Newman interview.

23. E. Kelly interview; Cody interview.

24. Johnnie Turner, interview by Brooks Lamb, July 9, 2015; Fred
    Davis, interview by Brooks Lamb, June 25, 2015. For more
    information on the segregation of the zoo and of Memphis in
    general, see Laurie B. Green, *Battling the Plantation Mentality:
    Memphis and the Black Freedom Struggle* (Chapel Hill:
    University of North Carolina Press, 2007).

25. Turner interview.

26. Johnny Cash, interview by Terry Gross, "Johnny Cash: In His
    Own Words," *Fresh Air,* National Public Radio, November 4, 1997.

27. Blanchard Tual, interview by Brooks Lamb, July 26, 2015.

28. Cohen interview; E. Kelly interview.

29. E. Kelly interview.

Chapter 2

1. Davis interview.

2. Turner interview.

3. Ibid. Turner states that there were seventeen students, includ-
   ing herself, who participated in the Youth for Christ sit-in at
   the Overton Park Shell. In his book *The Last Segregated Hour:
   The Memphis Kneel-Ins and the Campaign for Southern Church
   Desegregation* (New York: Oxford University Press, 2012),
   Stephen Haynes gives this number as fourteen students.

4. Turner interview.

5. *Watson v. City of Memphis,* 373 U.S. 534 (1963). It is important
   to note that the Memphis Zoo had integrated in 1960, three
   years prior to the *Watson* ruling. In chapter 8 of *Battling the
   Plantation Mentality,* historian Laurie B. Green explains that

the Memphis Committee on Community Relations worked
with city officials to negotiate the integration of the Memphis
Zoo in October 1960.

6. Turner interview.

7. Cody interview.

8. Basic statistics regarding the early years of the interstate con-
troversy were verified through an article written by Tannera
George Gibson: "Not in My Neighborhood: Memphis and the
Battle to Preserve Overton Park," *University of Memphis Law
Review* 41, no. 4 (2011): 725–43.

9. Newman interview.

10. Gigi Wischmeyer, interview by Brooks Lamb, July 8, 2015;
Jimmy Ogle, interview by Brooks Lamb, July 14, 2015.

11. Sally Jones Heinz, interview by Brooks Lamb, July 10, 2015.

12. A 1964 article on the front page of one of Memphis's most
prominent black newspapers advertised that "the purpose
of the rally is to protest the planned construction of U.S.
Interstate 40 expressway through the park. This expressway,
if built, would take . . . hundreds of trees from one of the few
virgin stands left in the Midsouth." Explaining that an "inter-
racial committee" had formed to help protect Overton Park,
the paper encouraged its readers to attend and support the
park. "Save Overton Park Rally Sunday, 3:30," *Memphis World,*
July 25, 1964, http://crossroadstofreedom.org.

Chapter 3

1. Newman interview; Department of Transportation Act of
1966, Pub. L. No. 89-670, 80 Stat. 931 (1966).

2. Summary judgment is a legal term that refers to a judgment

entered by a court in favor of one party and against another party without a full trial. A summary judgment may be issued either on the merits of an entire case or on certain points within that case.

3. Newman interview; *Citizens to Preserve Overton Park v. Volpe,* 401 U.S. 402 (1971).

4. The role of judicial review in *CPOP v. Volpe* is highlighted by many legal scholars and law textbooks. In his textbook *Administrative Law,* 4th ed. (Thousand Oaks, CA: SAGE Publications, 2006), Steven J. Cann specifically cites the Overton Park case to explain the role of judicial review as it pertains to decisions made by federal agencies. Christine B. Harrington and Lief H. Carter's textbook, *Administrative Law and Politics: Cases and Comments* (Washington, DC: CQ Press, 2015), also specifically discusses *CPOP v. Volpe* in regard to its impact on administrative law. *Environmental Law for Engineers and Geoscientists* (Boca Raton, FL: Lewis Publishers, 2002) by Robert Lee Aston, on the other hand, discusses the case's impact on environmental law.

5. Jesse H. Merrel, "Highway Robbery in Memphis—IV: Environmentalists Spilling Blood," *Transportation Topics,* May 1973.

6. Newman interview.

7. Willy Bearden, interview by Brooks Lamb, June 23, 2015.

8. Ibid. The famous album cover that Bearden references was for Trapeze's 1972 album, *You are the Music . . . We're Just the Band.*

9. Ibid.

10. Ibid.; Heinz interview.

11. Heinz interview.

12. Tual interview.

13. Wischmeyer interview.

14. Ogle interview.

15. Newman interview.

16. Bearden, *Images of America: Overton Park,* 120.

17. Martha Kelly, interview by Brooks Lamb, June 30, 2015; E. Kelly interview; Heinz interview; Wischmeyer interview.

18. Heinz interview.

19. Ibid.

Chapter 4

1. Janet Hooks, interview by Brooks Lamb, September 22, 2015.

2. Bearden interview.

3. Melanie White, interview by Brooks Lamb, July 13, 2015.

4. Wischmeyer interview.

5. Cody interview.

6. Ogle interview.

7. James Jalenak, interview by Brooks Lamb, July 13, 2015.

8. Kathy Fay and Richard Meek, interview by Brooks Lamb, July 15, 2015.

9. M. Kelly interview.

10. Bearden interview.

11. Donnie Bailey, interview by Brooks Lamb, November 12, 2017.

12. White interview.

13. Ogle interview.

14. Bearden interview.

Chapter 5

1. Cohen interview; Bearden interview.

2. Bearden interview.

3. Hooks interview.

4. Cody interview.

5. Jalenak interview.

6. Fay/Meek interview.

7. Jalenak interview.

8. Heinz interview.

9. M. Kelly interview.

10. White interview.

11. Tual interview.

12. Anne Pitts, interview by Brooks Lamb, July 30, 2015.

13. Bearden interview.

14. Tual interview.

15. The Citizens to Preserve Overton Park group that was formed in 2008 and still exists today is unaffiliated with the group from the twentieth century that stopped the interstate from destroying the park.

16. Newman interview.

17. Ogle interview.

18. Fay/Meek interview.

19. Jalenak interview.

20. Kelly interview.

21. White interview.

22. Newman interview.

23. Tual interview; Bearden interview.

Epilogue

1. Bearden interview.

2. Ogle interview; Hooks interview; Fay/Meek interview.

3. Bearden interview.

Page numbers in **boldface** *refer to illustrations.*